Self-Field Theory

A New Mathematical Description of Physics

Self-Field Theory

A New Mathematical Description of Physics

TONY FLEMING

PAN STANFORD PUBLISHING

Published by

Pan Stanford Publishing Pte. Ltd.
Penthouse Level, Suntec Tower 3
8 Temasek Boulevard
Singapore 038988

Email: editorial@panstanford.com
Web: www.panstanford.com

British Library Cataloguing-in-Publication Data
A catalogue record for this book is available from the British Library.

Self-Field Theory: A New Mathematical Description of Physics

ISBN 978-981-4303-00-2 (Hardcover)
ISBN 978-981-4303-01-9 (eBook)

Printed in the USA

Contents

Preface

If Maxwell and Lorentz had been informed that self-field theory (SFT) could solve the vector equations named in their honour as closed-form analytic expressions, they would not have been surprised as they had learned their mathematics before the advent of quantum theory. To them analytic closed-form solutions would serve to verify the physical basis of electromagnetics. On the other hand, if Heisenberg had been told the same, he would have been very surprised since to him uncertainty was a part of the fabric of reality, and his quantum world was indeed uncertain and never knowable beyond a level of uncertainty. If Einstein had been so informed he would have been delighted as he was unconvinced of the probabilistic nature of quantum theory. A unifying principle beyond uncertainty that would enable physics to be described in a deterministic fashion had eluded him during his later years. SFT's initial success was to solve the Maxwell–Lorentz (ML) equations for the hydrogen atom using bi-spinors to describe the electromagnetic (EM) motions of particles and fields in two orthogonal rotations. What this Bohr-like solution reveals is a previously unknown photonic layer of solution within the hydrogen atom that expands present physical knowledge. Via the mathematics of SFT an alternative and fresh description is available of the physics gained from quantum theory over the past 80 years.

It is said that those who do not learn the lessons of history are doomed to repeat its mistakes. In part this book is a concise history of classical electromagnetics and quantum field theory (QFT), especially where these theories are incomplete. We need to learn some of the mistakes of both to understand a new powerful EM formulation. If electromagnetics has thus far produced a technological bounty, then we can look towards a future with a more refined technology. This refinement will involve photonics, biophotonics, phononics, phonophotonics, nuclear physics, energy production, and a range of medical applications that include biophotonic manipulation of the DNA, to name a few. It is unclear how far the quiet revolution that is occurring in mathematical physics will take us. The climate change involving governments across the world demands a radical rethinking of our technological base — how to produce, how to get energy, where to live, and how to manage our transport needs. This small repository of new insights into our physical knowledge comes at a most propitious point in time and will allow us to rethink how to traffic goods and services across the globe. In addition this new knowledge may allow us novel methods of medical treatment. The expected lengthening of our life spans will demand changes to the basis of our industrialization. Older forms of industry need to be replaced by methods that incorporate new concepts.

Since the advent of the quantum theories that emerged before World War II, the theory of self-fields has remained a fossilized relic of classical EM theory. The work was initiated in 1903–04 by Abraham and Lorentz, who examined the self-field of a single particle, the electron. They studied the effect of its self-field on the electron itself by modeling it as a spherical surface of charge that was allowed to move at near-relativistic speeds. They found inconsistencies with their early form of self-field theory. It was their work at the start of the 20th century that alerted scientists, including Planck, Bohr, and others, to look for alternative methods of solution other than classical electromagnetics. It was not until 2005, when the self-fields were treated as *mutual* effects between two or more particles, that the mathematics finally made sense and the hydrogen atom was solved via a modified form of electromagnetics termed SFT.

SFT immediately impacted upon classical EM theory and quantum theory. SFT involves elastic collisions and a new concept of charge along with an understanding of the essential differences between continuum and quantum physics. Early in the gestation period of SFT the author realised that it was a formulation that had application across physics. It occurred to him that SFT was perhaps the unified field theory that Einstein had unsuccessfully searched for in his later years. Several of Einstein's statements indicate that like the author he was not convinced of the probabilistic foundations of QFT. He alluded to God not playing dice, indicating a belief in a certain level of determinism within physics, a prominent feature of SFT compared with QFT, that is to say the current pre-SFT implementations of QFT. Einstein was right, yet statistical information at least provided partial understanding in the interim until SFT resurfaced in a modern form.

It also occurred to the author that such models of reality as SFT, and QFT, are just that: models. In mathematics the difference between a model and a theory is the extent to which a model matches experiment. To an applied mathematician, a mathematical model needs to be tested against physical evidence, experiment, measurement, collected data of some kind. There are regions of applicability and other regions of inapplicability. In the mathematics of Descartes, the 17th-century leader of the rationalists who opposed the empiricists, the mind has primacy over corporeal matter: "I think, therefore I am." But mathematical models, similar to musical compositions, can be wrong. Beethoven, Mozart, and Bach all understood this intimately. The mind is not the essential core of living, sentient existence; something else, call it the soul, directs the mind and the body in its creative role. Hence SFT, QFT, or any other mathematical model of the physics is inaccurate to some degree. Mental process cannot replace the actual physics.

Heisenberg's uncertainty is similarly involved in this discussion, but on the other side of the ledger. Heisenberg believed that uncertainty was a part of the fabric of reality; we must simply learn to live with it. No one could look at the photon without disturbing it. However, the truth is, we humans have the ability to use our intellects to examine the insides of the photon without disturbing it. Heisenberg's comment was an insight into experimentation as he saw it in the 1920s. Hence science must walk a gangplank across the mental chasm between ignorance and knowledge. To the author, knowledge like the fields of SFT is fractal, as is consciousness. We learn layers of a knowledge onion always approaching a more refined, more encompassing view of reality. But equations are just a model of the physics and not the driver of the physics. Like a painting or a photograph, they are an *image* only and do not incorporate the entirety of physical existence. This is not to denigrate the image so conceived; the use of virtual reality to assist surgeons to perform ever more delicate operations is testament to the power of image. Our intellects have an innate ability to perceive the physics, and we have learnt to write these perceptions as mathematical equations. We can see that we are after all made in God's image, a faulty being maybe, but a wonderfully gifted and highly evolved creation without doubt. We have an earnest duty of stewardship over the Earth and its amazing range of animal and biological creatures. Buzz Aldrin, who along with Neil Armstrong, was first to set foot on the Moon, believes that it may be our role to explore and propagate the universe. He too may be right.

Many distinct areas of physics are covered within the book. This has necessitated a single view regarding equations and the form of their application. With regard to notation and convention used within the book, we will in our formulation use the speed of light c and Planck's constant h rather than the current convention in high-energy physics to normalize $c = 1$ and $h = 1$. We do this because a great deal of important physics can be uncovered by noting the changes to c and h across different particles such as the electron and the photon. Indeed both c and h are seen as media-dependent parameters rather than constants. This should not be seen to contradict present understanding but rather to allow expansion into novel regions of applicability. Often a physical situation can be explained in terms of the permittivity ε and permeability μ of different regions inside atoms, and molecules, as both c and h depend on these constitutive parameters. These material parameters control the energy density within various EM regions, and in part within the strong nuclear regions where quarks and gluons move in dynamic equilibrium.

Modern applications of electromagnetics, such as photonic crystals and biophotonics, include generalized materials or regions that necessitate a more complete EM analysis in which ε, μ, and sometimes the conductivity σ must be included. At lower-energy levels such changes to the media involve the external energy–frequency $(E = h\upsilon)$ response of photons, for example within the ionosphere as photons approach the Earth on transit from the Sun. While the frequency of an impinging particle is maintained, its wavelength may change along with its self-energy structure. The Compton wavelengths can be viewed within SFT as an effect upon the internal self-field structure of photons and electrons, while pair production processes are seen as ionization levels of bonded particle pairs with annihilations — the inverse process particle pair bonding.

Another mathematical usage requires explanation. In QFT the term *spinor* denotes a normalized element of complex vector space used to aid the algebra of the quantum solution. In SFT the term spinor is a physical entity that, for instance, may describe the motion of a particle as a rotating vector of variable length. Such motions are time-varying EM solutions to the ML equations, as detailed in Chapter 2. Hence they are a basis for the general solution to the ML equations when used as a set of eigenequations. As with c and h, the physical spinors reflect the actual size of a particle's motion.

As the quantum methods evolved, the terminology that evolved was naturally associated with the emerging contemporary physical understanding of an effect. While this process is unavoidable, it does not provide a cohesive overall picture of the terms and results in a somewhat piecemeal nomenclature. One such instance is quantum spin S and its associated number s. Quantum spin is related to the theories, e.g. Bohr theory, that came to light in the early days of quantum mathematics and atomic chemistry. In SFT, spin refers to the cyclotron effect between two or more particles, somewhat different to the concept of quantum spin as a discrete quantum number or set of numbers. In SFT spin is often not an intrinsic quality, but an extrinsic mutual effect; indeed it can be both. An example of this are the Solar System and galactic spins of our own Sun, and we shall examine a new hypothetical form of gravity at the galactic level in Chapter 4. Similar terminological differences exist between high-energy physics and SFT as it applies to the strong nuclear region. Quantum spin, isospin, charm, strange, up, and down are all associated with their tri-spinorial counterparts: the orbital, cyclotron, and nuclear rotations and their frequencies and radial distances from the centres of rotation.

Within the book, except for comparison, we do not use Lagrangian methods but remain with the ML equations, what is called the Maxwellian. Although the Lagrangian formulation is often considered a more concise and elegant form of mathematics, the necessity for considering gauge conditions associated with raising the order of the equations to the second-order wave equations more than offsets any presumed analytic disadvantages of remaining with the first-order curl and divergence equations. In this regard comparison between QFT and SFT is similar to the comparisons known to numerical analysts between integral and differential equations. While the former, such as the use of Green's function analysis known within the physics community as Feynman diagrams, can be used for problems where regions are homogeneous and isotropic, differential formulations such as the ML equations are more suited in regions of varying inhomogeneity, including the interiors of atoms. This may also apply to cosmology where the cosmological principle of an isotropic and homogeneous universe may well prove to be inaccurate.

At this early stage of the development of SFT the author has been traveling widely around the globe in an attempt to reach as wide an audience of physicists, biophysicists, and mathematicians as possible. As of the end of July 2010 he has clocked up around half a million kilometers in air travel to the United States, China, Russia, Europe, Asia, and numerous other locations and is well and truly "traveled out." The task of peer review is a long and arduous process, and as regards SFT it is not complete. As Churchill said, "Now this is not the end. It is not even the beginning of the end. But it is, perhaps, the end of the beginning." Speech at the height of WWII, November 1942. Perhaps in its evolution as a major development in science it is not even the end of the beginning but a period just after the middle of the end of the beginning (with all due reverence to Sir Winston). Thus much of SFT is hypothesis remaining to be experimentally verified.

That said there is much important evidence to support SFT as a fundamental mathematical basis for physics. This includes evidence contained within the formulation itself. Planck's constant is an experimentally based number obtained by Planck in his blackbody calculations and Einstein in his photoelectric experiments, both of whose efforts prefaced quantum theory. Quantum theory was based on a heuristic formulation that included Planck's constant. On the other hand the SFT formulation inherently includes a term the author has called Planck's "number," which corresponds to Planck's constant to 7 decimal places of accuracy. SFT is thus seen to be more fundamental than quantum theory. It can also be seen to a mathematical level that SFT corresponds to some extent to quantum theory. Hence we can have a certain degree of confidence that SFT will provide experimental correspondence in the same way to quantum theory. We can anticipate that the SFT formulation will be a more accurate match to the physics because it is inherently a more accurate model of the photon and the EM fields. That it is an improvement on classical EM (CEM) theory is also evident, mainly because it does provide a solution to the hydrogen atom. In fact SFT provides a "complete Bohr theory" solution, one that includes the magnetic effect.

In the Appendices a section called FAQ has been included to assist the student in the task of comparison between quantum theory and SFT. Many questions have arisen in the conferences and congresses the author has attended in the task of presenting SFT for peer review in the many areas of physics it has to date been so presented. We can anticipate that there will be numerous scientific disciplines where the scientific method will be enabled. Many "hard sciences" will be enjoined by other "softer" sciences where validation will eventuate. It may be that some of these softer areas have unreasonably earned a bad name, being called "pseudoscience" by self-proclaimed experts and skeptics. While it is not the author's aim to cause a reversion back to earlier times when charlatans were able to convince many by force of character, it is an aim to review science afresh without prejudice. This review is needed especially where medical science is involved. It appears to the author and his colleague Elizabeth Colorio that some forms of medical therapy may have a valid scientific basis, including acupuncture and the various forms of "frequency medicine," where different frequencies of interaction are being used for different organs and tissues of the body. This important part of their work is addressed in a separate book called *Inside the Photon: A Journey to Health*, also published by Pan Stanford Publishing.

The author would like to thank his colleague Elizabeth Bauer for her input as the one who was there during the process of distillation of the SFT and with whom the author could reflect ideas and fly balloons to test the proverbial waters via innumerable emails from Australia to various parts of the United States. The author is indebted to the editorial staff of Pan Stanford for their assistance in the production of this book. He would also like to give credit to the numerous illustrators whose work has been donated to the repository of Wikimedia Commons, in general a public domain collection of various media maintained by Wikipedia, the Free Encyclopedia. Wherever the term "Credit: Wikipedia" is given in a figure caption, the diagram, drawing, or picture has come from Wikimedia Commons, for which the author is very grateful. Parts of Chapter 2 are reprinted by permission from "Electromagnetic Self-Field Theory and Its Application to the Hydrogen Atom," Anthony H. J. Fleming, *Physics Essays*, **18**, 265 (2005).

Chapter 1

Introduction

Electrical and magnetic fields have been known in mathematical form since the laws of Coulomb and Ampere were discovered in the late 17th and early 18th centuries. Applying to macroscopic domains, modern atomic physics was in its infancy at that time. Both laws assume charge separation in a single nonlinear function, the square root of the sum of the squares of orthogonal distances involved in their separation. Within atomic and molecular theory the Pythagorean concept of distance has been utilized by both classical electromagnetics and quantum theory. Einstein's relativity gave the first hint that in some phenomena, separations in orthogonal directions do not couple but stay separate. Thus electromagnetic fields in atoms consist of two fields each causing the atomic particles to rotate in orthogonal planes. If charge separation includes both centres of rotation, electromagnetics can analytically solve for the motions. In this chapter we examine how classical electromagnetics failed at the atomic level and quantum theories were deemed necessary, thus dominating 20th-century physics. The classical fields were responsible for this historical failure. A selective sweep across scientific knowledge provides a preview of self-field theory.

Self-Field Theory: A New Mathematical Description of Physics
By Tony Fleming

www.panstanford.com

1.1 Classical Origins

Pythagoras, Plato, and Archimedes are all to some degree associated with the recent progress in self-field theory (SFT). Pythagoras was the first to use regular solids to fill space. He used triangles or tessellations, or more generally polygons in 2-D space and polyhedra in 3-D space. There were five regular polyhedral solids associated with the physical world. Plato recorded this system of forms in his writings. SFT revives this concept in regular forms or splines that can be associated with diverse physical and biological entities such as the ionosphere, snowflakes, worms, shells, and DNA. Archimedes used polygons that both inscribed and circumscribed a circle to estimate π. This is an early form of finite element method, the numerical method of choice of modern engineers and mathematicians. Interestingly, integral equation methods similar to the finite element method are used exclusively to solve the equations associated with quantum field theories (QFTs) of quantum mechanics (QM), quantum electrodynamics (QED), and quantum chromodynamics (QCD). Perhaps of all ancient mathematics, the theorem of Pythagoras stands at the very pinnacle. Schoolchildren perhaps since these ancient times have learnt that "the square on the hypotenuse equals the sum of the squares on the other two sides". This concept of shortest distance is intimately associated with solving physical equations as in modern physics. However, we need to carefully examine its use within the eigenvalue problems that are solved via QFTs and SFT.

What is implicitly understood is that Pythagoras was talking about geometric distances known to him. He was assuming he could measure or walk along the hypotenuse of any triangle he wished (Fig. 1.1). He was not talking about charged particles such as the electron and the proton moving through atomic distances or light moving through cosmological distances. Taking an example Pythagoras would understand, we might ask him to walk the shortest distance between Samos, where he was born, and Athens, the main city of ancient Greece at that time, ca. 500 BC. He would tell us that since Samos is an island he could not possibly walk between the two locations. With a boat he could sail the shortest path. Thus distance depends on the physics and how many forces are applied.

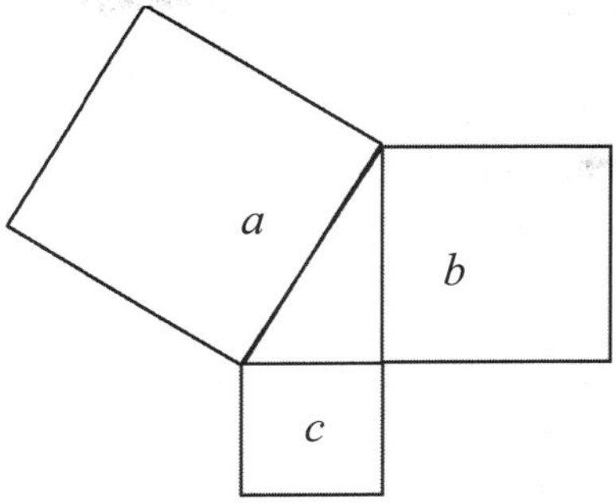

Figure 1.1. Pythagorean distance: $a^2 = b^2 + c^2$ a, b, and $c \in \mathbb{R}$.

Similarly Coulomb used the torsion balance (Fig. 1.2) to detect the charges between small metal-covered balls suspended by silk threads. In his equation for the electrostatic force between the balls, he was again assuming macroscopic distances directly measurable by him and commensurate with the size of the distances between them. His use of the balance combined the vertical or

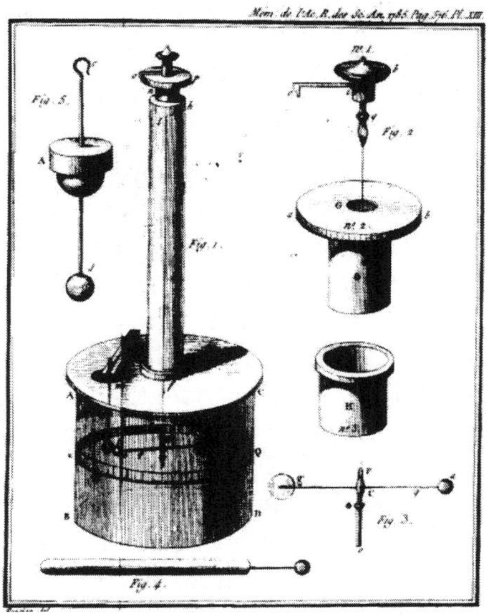

Figure 1.2. Coulomb's torsion balance. (Credit: Wikimedia.)

spherically radial component of gravitation and the horizontal, spherically tangential component of the electrostatic force that were measured along the hypotenuse of the silk string. Pythagorean distances were again implicit in Coulomb's equation. However, as is now known, the quantum world is broken up into discrete distances and times in its description of space-time. Pythagorean distance may be a good approximation for macroscopic domains, but we need to be careful in using it for microscopic, nanoscopic, and femtometre domains as is necessary for studying atomic and nuclear systems, or indeed for cosmological domains such as Newton's gravitational law. The reason why the early electrostatic and magnetostatic laws were uncoupled electromagnetic (EM) effects is that either the electric (E) or magnetic (H) field dominated in a particular instance, and this dominance can occur in any domain Theorists often seek out such simplifying circumstances to make the mathematics easier.

Much of the modern history of electromagnetics is connected to the history of relativity and quantum physics. The debate between Einstein, Heisenberg, and Bohr mainly concerned the role of probabilities and continued unabated for decades during their lives and still echoes today. The arguments pivot around the deterministic or probabilistic nature of physics; in other words, causality, cause, and effect. As with the wave or particle nature of light, both views are partly correct; nature does employ probabilities, but this is done within a deterministic physics. According to SFT a particular physics is quantum or continuous depending on the ratio of the masses and energies involved in the interactions. Planck in 1900 discovered that discrete frequencies must be involved in his law of blackbody radiation. In terms of atomic physics Bohr combined the quantization of angular momentum and the classical theory of the hydrogen atom to reveal a model that accurately predicted the wavelengths and energies associated with the hydrogen atom (Fig. 1.3). In this model, wavelength is given by the De Broglie equation $\lambda = \frac{h}{m_e v_e}$ where h is Planck's constant, v_e is the velocity, and m_e the mass of the electron. These frequencies point to the fact that the atom behaves like a resonant system. As we approach the ground state frequency for the hydrogen atom, the *grainy* or quantum nature of the physics starts to appear. The wavelength of

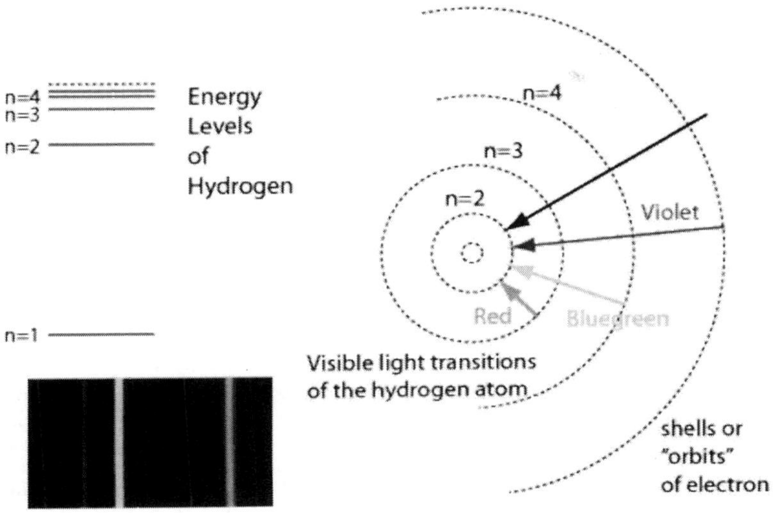

Figure 1.3. Quantized levels of the atom. (Credit: Wikimedia.) See also Color Insert.

the electron $\frac{h}{m_e v_e} = \frac{6.62\times10^{-34} J\ s^{-1}}{9.11\times10^{-31}\text{kg}\times1.9\times10^{6}\text{m s}^{-1}}$ is estimated as 7 Å, not macroscopic, or even microscopic, but atomic.

The distances are atomic in scale and the physics is quantum; any multiple of the principal wavelength or principal frequency is also resonant. Euler and Lagrange first investigated the roots of differential equations as a systematic study in their own right. By decomposing functions into discrete series of continuous functions, Fourier breached the divide between the quantum and continuous worlds. The mathematics of resonance was known in the early to mid 19th century. Faraday had performed experiments pointing to the quantization of charge in this period. Hence both the mathematics and physics of quantum phenomena were well understood long before the quantum era post 1927.

In terms of the atom there were other important mathematical concepts known by the 1850s. Euler studying the rotations of rigid bodies determined the importance of the principal axes (Fig. 1.4). Cauchy, who also theoretically studied the roots of polynomials, detected a spectral nature behind multi-resonant structures, including the characteristic root now known as the

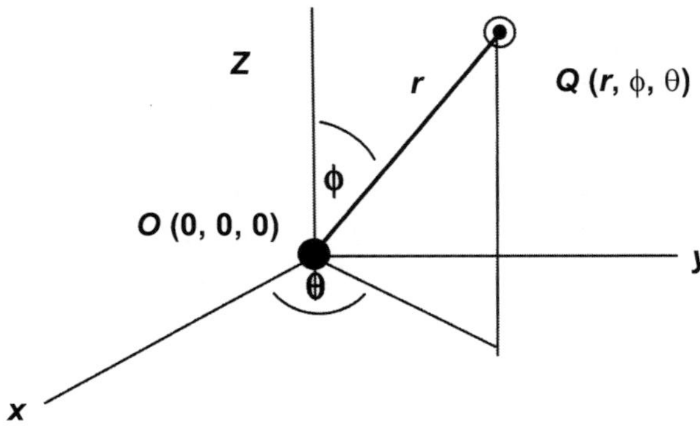

Figure 1.4. Coordinate system for the hydrogen atom using spherical coordinates.

principal eigenvalue. Lagrange associated the eigenvalue solutions with particular eigenvectors, including the inertial coordinate system in which the system was in balance. The final key to analytically solve differential equations is to separate the variables into the coordinates of the principal eigenvector. Laplace and Fourier found orthogonal series linked to the eigenvalues and eigenvectors.

Maxwell was the first to understand the EM plane wave. He saw that there were *two* interacting fields causing *two* forces to act upon charged particles with *two* orthogonal motions. While Coulomb's and Ampere's laws concerned single fields, the EM plane wave showed that EM fields occurred in two orthogonal directions and therefore for charged particles in relative motion two orthogonal motions follow.

Using separation of variables Hertz in 1888 developed a mathematical formulation based on potential theory for the dipole antenna that explained how radiation came about. In this case it is noted that a phase difference of $\pi/2$ between the E- and H-field energy densities in the radiation zone gives a zero radiation condition. In 1908 Mei found an analytic solution to scattering by layered dielectric spheres where each layer has a different value of relative permittivity. Different forms of the Bessel functions for the spherical EM fields are needed for the near fields and the far fields. Two fundamentally different types of field were needed to obtain a zero sum balance in the radiation region; one represented an outgoing traveling wave, while the other represented an inward traveling wave. Hence scattering brings about an open standing wave around the sphere.

For the hydrogen atom, the principal wavefunction obtained by the method of Shrödinger's wave mechanics using separation of variables in spherical coordinates is $\psi_1 = \frac{1}{\pi^{1/2}}\left(\frac{q}{a}\right)^{3/2} e^{-Zr/a}$. Where the atom is in equilibrium the application of the separation of variables to the EM forces should have two distances in its functional form and not the single distance of Shrödinger's wave function. This lack of two fields, both the E *and* the H fields, affects Bohr theory in a similar fashion. Bohr theory is incomplete because it does not include any H effect upon the electron. If Bohr had included a magnetic rotation he may have followed down the theoretical path of SFT treating both the electron and the proton with two coupled fields. The two particles within the hydrogen atom interact via their EM self-fields consisting of both E- and H-field components. Only by including *both* fields can an equilibrium or resonance condition be analytically determined. The quantum method relies on the experimental observation of specific frequencies matched to the eigenvalue formulation. If only one field is included as in the classical and quantum field forms, then the methods fail to a greater or lesser degree. This is the source of an inaccuracy known in quantum theory as *uncertainty.*

The unit circle and the Argand or complex plane have a special place in classical mathematics, and they play an important role in the SFT solutions for the EM motions within the atom via the eigenvalue method. *Imaginary* numbers have been known at least since Cardan encountered them in studying the roots of cubic polynomials in 1545. Such numbers and their Pythagorean interpretation have an implicit link to the roots and determinants of matrices that are encountered in the mathematics of multi-resonant structures. In SFT a rotating vector is termed a spinor, defined as $\sigma(r, \omega) = re^{e^{j\omega t}}$ (Fig. 1.5), where time is a parametric variable. Here the spinor rotates in an orthogonal direction to motion. For a particle stationary in the direction orthogonal to the Argand plane an observer might see the particle moving around the circle or see a particle moving up and down in time depending on the observation point. Similarly, for a particle moving in the orthogonal direction,

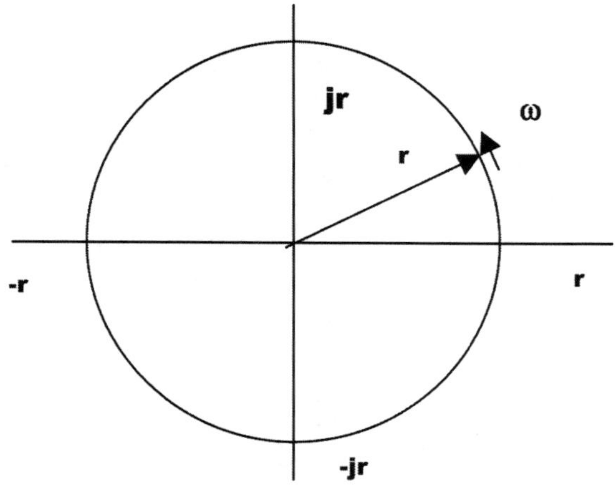

Figure 1.5. Spinor σ rotating at angular speed ω around a circle of radius r in an Argand plane.

an observer might see a particle moving along a helical motion or moving up and down as a sinusoid depending on the relative observation point. By varying the parametric values of the spinor it is thus possible to create spirals, vortices, helices, traveling waves, standing waves, etc. Hence the motions of EM fields can be treated via these physical spinors (Fig. 1.5).

The spinors defined above are very different from the spinors known to group theory that are related to the determinants of unitary matrices (det $=$ 1). Although the application of spinors to field theory is a relatively recent study the origins of group theory lie in the early algebraic attempts to solve polynomials and differential equations. Groups are found in number theory,[1] algebra, and geometry and also occur where symmetry is considered, a discipline of much concern in numerical methods. The group $O(2)$, orthogonal 2×2 matrices define rotations in the Argand plane, the unit circle termed SO(2). It is well known that EM fields, both E and H fields, induce $O(2)$ motions on moving charges. While the Standard

[1]Although Fermat is outside this study of the hydrogen atom, he stands as the greatest of all number theorists, and without his "dabblings" (he was a practicing lawyer) the world of physics would be a much duller and primitive place. Fermat corresponded with Pascal, who is the modern progenitor of the Binomial theorem, the opus at the heart of number theory and probability theory.

Model uses more complex groups to restrict the types of fields that can occur in the EM, strong, and weak nuclear interactions, SFT uses $O(2 \times 2)$ or $O(4)$, a bi-spinorial field that accounts for the rotational motions caused by both the electric and the magnetic fields of any charged particle such as electrons or protons. These $O(4)$ matrices are diagonal and completely orthogonal, unlike conventional matrices, where off-diagonal terms occur. Thus the eigensolutions of the hydrogen atom are analytic series similar to Mei scattering. Multi-electron atoms and molecules can also be solved as analytic series. This is a fundamental difference to the numerical approach of quantum theory. The strong field can be similarly considered to consist of *three* field particles or *three* fields causing *three* physical rotations, i.e. $O(3 \times 2)$ or $O(6)$.

Thus the EM motions involved do *not* involve Pythagorean distances, but pairs of orthogonal rotations in space-time. In the case of the shortest distance between Samos and Athens, Pythagoras assumed he could walk or sail in the precise direction of the shortest path. He could in the macroscopic terrain get his muscles to move his body, within a certain accuracy, to walk along the shortest direction and, if adept enough, sail along this shortest direction. Similarly, in his experiment, Coulomb assumed he could choose the separations, within a degree of accuracy in any macroscopic orientation and at any radial separation he desired. However, in the atom Bohr could *not* choose any separation, only distances behaving like atoms. The allowable radial distances and frequencies form a grid in spherical space-time. Anyone familiar with drawing on a computer screen will be aware of the "snap to grid" function that separates the plane of the monitor into a grid. Similarly the forces within atoms cannot align in any direction or length but apply relative to the orientation of the moving particles. Certain minimum energy states and standing wave conditions apply. Planck's blackbody does not radiate *any* wavelength or frequency, only discrete wavelengths and frequencies obeying the laws of EM physics governing its motions. In brief we find that by assuming a bi-spinorial form for EM fields $r(r_o, \omega_o, r_c, \omega_c) = r_o e^{j\omega_o t} + r_c e^{j\omega_c t}$ we can solve for atomic motions.

Other examples of discrete space-times occur in physics. In acoustics (Fig. 1.6) the modes of a vibrating string are dependent on the physical parameters involved with the string, including its

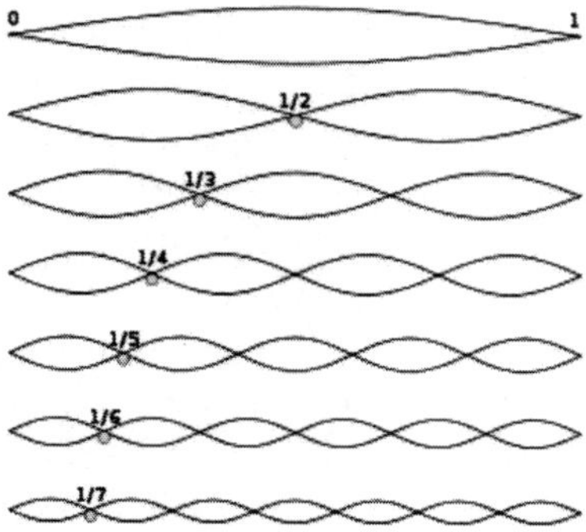

Figure 1.6. Modes of a vibrating string. (Credit: Wikimedia.)

length and its internal structure and how hard it is plucked. Standing waves only appear at certain distances, the eigensolutions, along the string depending on these physical parameters. In acoustics the parameters can be changed so as to keep the standing waves within the macroscopic domain. Hence if we change the length of a musical instrument, such as using the frets of a guitar, the physics is macroscopic and we can hear the quantum shifts as different notes. The same principles apply to electromagnetics. In some crystals both phonons, the quantum of sonar energy, and photons form a 3-D grid of "allowed" directions similar to a laser.

Composite structures of the photon and the phonon point to their unification in the form of the gluon. $O_{\text{A}}(2)+O_{\text{EM}}(4) = O_{\text{SN}}(6)$ where A is an acoustic spinor, EM is an electromagnetic bi-spinor, and SN is a strong nuclear tri-spinor. Thus the phonon is represented by an $O(2)$ matrix that can interact with the $O(4)$ matrix of the photon to form the $O(6)$ matrix of the gluon, the field particle of the strong nuclear force. These SFT groups of complex exponentials in $O(2)$, $O(4)$, and $O(6)$ are based on the classical eigensolutions used to solve polynomial and differential equations several centuries ago. Mathematicians from the times of Euler have been using complex exponentials to solve linear differential equations. Such usage is called empirical, where experience forms the basis for solution and is retained until experience suggests further modification. Quantum theory is based on empiricism. The scientific method too is based on empirical trial and error, the testing of hypothesis.

SFT points to a logical construction of field particles or bosons called *photon chemistry*, in which the photon, the phonon, and the gluon are all interacting elements of a group of spinors of different degrees of freedom. This is a subdivision of sub-photonic particles and their states similar to the structure of atoms and their states into logical arrays that has evolved from Mendeleev's 2-D periodic table of chemical elements to the 4-D array given by quantum theory and finally now to a 6-D array based on the composite photon and its family of compounds. While God may not play dice, the photon states can cause the Sun and its planets to flip their magnetic poles in a seemingly random manner.

A 6-D or $O(6)$ group is consistent with the equipartition of energy, for instance within diatomic gases into different "modes" as follows. The translational, rotational, or vibrational modes are excited and contribute to the heat capacity in different temperature ranges. At low temperatures only translational modes are excited and are the only modes available to the specific heat $PV = 3/2\,nRT$. At mid-temperature range the rotational modes come into play, so $PV = 5/2\,nRT$. Finally vibrational modes become available above a critical temperature, so $PV = 7/2\,nRT$ at normal pressure P, volume V, temperature T and where R is the gas constant $R = 8.31\,\mathrm{J}\cdot(\mathrm{mol}\,^\circ\mathrm{K})^{-1}$.

Newton's gravitational, Coulomb's electrostatic, and Ampere's magnetostatic laws all assume that the field covers all solid angles over $(2\pi)^2$ out from the source. SFT sees the field particles, the bosons, as stream-like, covering only a small beam of solid angle. The directionality of the EM field may be involved in the cosmological light that can and cannot be seen from Earth. In cosmology a current mystery surrounds the issue of *dark energy*, a form of energy proposed to be responsible for the accelerating expansion of the universe. Dark energy is thought to consist of over 70% of the mass in the universe. It may be that higher forms of radiation may, like the photon within SFT, be stream-like within the structures of the universe. A similar mystery concerns a form of matter known as *dark matter*, proposed to explain the rotational speeds of galaxies, orbital velocities of cluster galaxies, and gravitational lensing by galaxy clusters. All these effects suggest that galaxies, cluster galaxies, and the universe contain more mass than can be seen by us on Earth or by the orbiting Hubble telescope. One explanation may be that galactic radiation is gluonic and thus has more structural "fibre" than the form of gravity known to us at the solar system level. Structural degrees of freedom of the various bosons within the universe may inflate the apparent mass assumed equivalent to that involved in the Solar System.

It is here that we need to introduce another classical connection with SFT. This has relation to more generalized solutions to polynomials and differential equations than complex polynomials consisting of geometric circles. Archimedes, working on a line of classical enquiry that goes back to the origins of classical number theory, sought to estimate the geometric constant known as π, the ratio of the radius and the circumference of a circle, in fact any circle. Early forms of approximation are known to have been put forward by Egyptian, Babylonian, Indian, Chinese, and Greek mathematicians from about 1900 BC, including a clear reference to the approximate value $\pi \simeq 3$ within the Old Testament. Archimedes iteratively found a more accurate value for π by circumscribing and inscribing a given circle by two n-sided polygons (Fig. 1.7). He used a 96-sided polygon to estimate π.

In seeking a mass of the photon consistent with the results of QED, we shall see in Chapter 3 that according to SFT there is a *photon transition frequency* ω_γ. This parameter is the number of times a photon transits back and forth between the electron and the proton during a single cycle of the electron and the proton within the hydrogen atom. Using the exceptionally accurate experimental

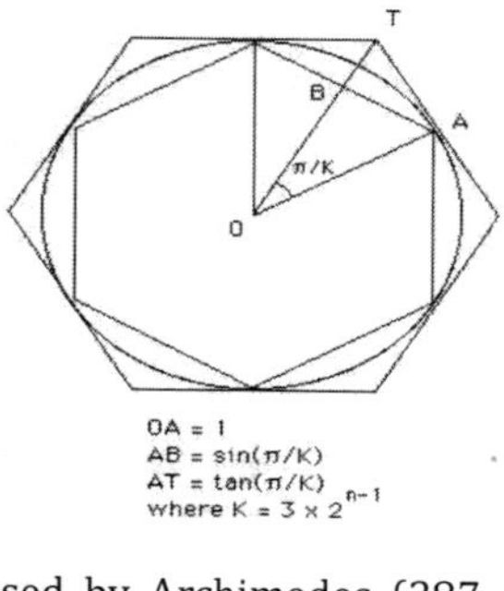

Figure 1.7. Method used by Archimedes (287–212 BC) to estimate π. (Credit: Wikimedia.)

and theoretical value of the Landé g-factor for the electron, a parameter used mainly within the confines of QED, we find that ω_γ approximates to the integer value 54. Thus we can equate Planck's quantum of energy with the binding energy of a photon of mass m_γ moving with the speed of light, c. Hence, while Archimedes used a 96-sided polygon, it seems that a quantum solution to the hydrogen atom involves a 54-sided polygon.

Thus a more complete solution to the characteristic root of the differential equation for the atom does not involve exponentials but rather piecewise splines. This piecewise nature of the actual physics agrees with the concept of coherent elastic collisions between the photon and the electron and the photon and the proton, which reveals a possible connection to how we might regard charge in terms of coherent collisions that sounds very familiar to the kinetic theory of gases and the statistical mechanics that led to QM. The Maxwell–Boltzmann statistical distribution is $\frac{N_i}{N} = \frac{g_i}{e^{(\varepsilon_i - \mu)/kT}}$ where N_i is the number of particles in state I, ε_i is the energy of the ith state, g_i is the degeneracy of energy level i, the number of states with energy ε_i, μ is the chemical potential, k is Boltzmann's constant, T is absolute temperature, and Z is the partition function. This distribution (Fig. 1.8) was the basis for Planck's work in blackbody radiation. Hence there is a consistency between the classical EM, quantum, and self-field theories in terms of elastic, coherent collisions between bosons and matter.

In photonic phenomena including the distances traveled by light or by particles undergoing atomic EM forces that continue in dynamic equilibria, the separations of standing waves are not established under the laws governing Pythagorean distances but via bidirectional rotations. Dynamic forces can be applied only in certain spherical directions. Thus E and H fields must not be calculated using Pythagorean distances inside atoms (Fig. 1.9). As with Planck's investigations of blackbody radiation where frequencies were found to be precise multiples of what are termed Planck's constant, only certain directions and distances, and their multiples, are allowed. The electron and the proton do *not* interact in such a way as to move along the hypotenuse of a triangle. The motions of electrons and protons in atoms due to their potential and kinetic energies do *not* interact. It is this essential insight into atomic physics that enables

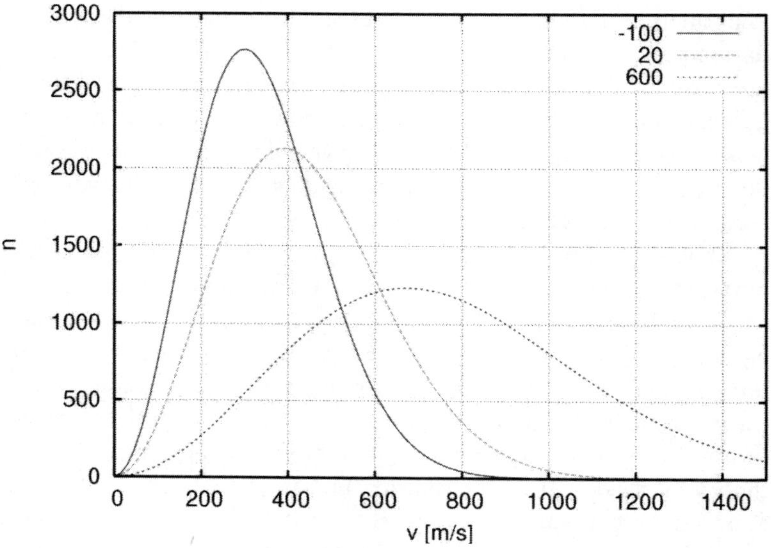

Figure 1.8. Maxwell–Boltzmann distribution for thermodynamics. (Credit: Wikimedia.) See also Color Insert.

SFT to analytically solve for these parametric entities. Similarly it is the failure of conventional QM and QFTs to grasp this fundamental behaviour that led them to remain with only a numerical legacy of nearly all studies within the 20th century.

Thus the atomic phenomena associated with balanced EM forces are eigenvalue problems, similar to the vibrating string shown in Fig. 1.6. Each solution of the overall atomic eigenvalue problem is an orthogonal solution in space-time in the same way that a general solution to a differential equation can be composed of a number

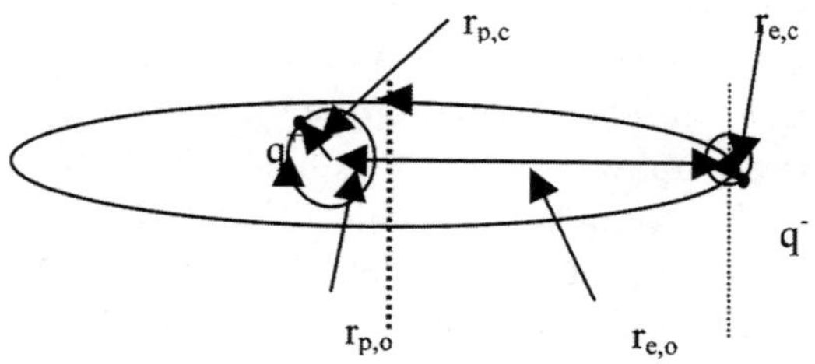

Figure 1.9. Atomic separations $\mathbf{r}_e = \mathbf{r}_{e,o} + \mathbf{r}_{e,c}$.

of particular solutions that are independent of each other. Each particular solution exists in isolation to other possible particular solutions. It is well known that atomic systems can be posed as eigenvalue problems having general and particular solutions. The heuristic method behind quantum theories contains the experimentally determined Planck's constant[2] that appears as an analytic variable of motion within the SFT formulation.

The various elemental atoms are resonant structures; each electron in its own shell has been added via some cosmological process in the past as if it were an eigenvalue problem where an eigensolution has been added to a layered configuration, structured like an onion. This history of successive eigensolutions gives a record of the ambient energy of these elements at previous points in time. Although these concepts are known from the various QFTs and physics in general, SFT yields closed-form solutions for the complete atom without the need for numerical methods. Exact solutions for atomic and nuclear physics via SFT make atomic chemistry more precise, and science and engineering more tractable. At the same time strong classical links are inherent within the SFT formulation.

1.2 1671–1903

Electromagnetics has been studied since classical times in the form of light. Since Newton and Young one intriguing question has been whether light is a wave, a particle, or both. Newton felt that refraction through a prism, like particles moving horizontally under gravitation as studied by Galileo, could be explained as different miniscule particles going through the prism at different speeds. Equally, Young's two-slit experiment could be related to wave theory. E and H fields per se were not understood mathematically until Coulomb first put forward his law of electrostatics just prior to the French Revolution using a torsion balance to measure the forces

[2]In the SFT formulation Planck's "constant" appears as a variable of motion depending on electron speed. This SFT parameter will be referred to in this book as Planck's number.

between weak charges. Using similar triangles the forces of repulsion or attraction due to various charges on two small metallic balls was related to the gravitational force and found to be an inverse function of distance squared. Thus the force of electrostatics was found to obey a similar law to Newton's gravitational law. Just prior to Coulomb's experiments, the explorer James Cook with the aid of an accurate marine chronometer had discovered Australia, Antarctica, and New Zealand. Longitude had been resolved as a matter of accurate timekeeping. The difference between local time at high noon and an accurate measure of total time was known to be capable of providing the longitude. From 1737 to 1765 Harrison, a clockmaker, invented his marine chronometer counteracting the effects of gravity and the motions of the ship. Coulomb's torsion balance was an extension of the concept behind Harrison's chronometer. At that time, while magnets were known to indicate north–south, geomagnetism was not well understood and in practice the method was inaccurate near true north, leading to many ancient ship wrecks. In 1820 Oersted demonstrated the deflection of a compass needle near a current carrying wire. Ampere found a mathematical function for the induced force between two current-carrying wires of similar form to Coulomb's electrostatic law.

In 1831 Faraday discovered EM rotations and EM induction indicating an intimate connection between E and H fields. He saw that charge measurement depended on the electrical state of the observer. To Faraday the electrical state, the charge of a particle, was not a fluid, as then thought, but depended on magnetic forces. Thus he established the field theory of electromagnetism. Applying his efforts to Faraday's findings, Maxwell discovered the four mathematical equations that define electromagnetics in terms of their wave nature. The 20 electrodynamics equations he had discovered during his life were posthumously reformulated down to four by Heaviside. His work heralded Einstein's theory of relativity in 1905. Maxwell also contributed to the kinetic theory of gases and the statistical methods that were to play a major role in the quantum theories of the 20th century. His work output was huge and diverse and gave impetus to a unified view of physics, after which Einstein was to strive unsuccessfully in his later life.

Searching after the electric and magnetic unity that Faraday perceived some decades earlier, Maxwell's theoretical and experimental findings that light and electromagnetics had the same speed to the limits of 19th-century experimental techniques led to the unification of optical, electrical, and magnetic effects into the one phenomenon of EM waves. He predicted that EM waves propagate in free space with speed c, where $c^2 = \frac{1}{\mu_0 \varepsilon_0}$, μ_0 is the magnetic permeability of free space and ε_0 is the electric permittivity of free space. In addition, light was known to act as a transverse wave, as was the EM field. Although it took a while for the scientific community to convince itself of his propagating waves in free space, it was soon obvious Maxwell was correct.

In 1888 Hertz confirmed the existence of EM radio waves using an early form of wire antenna, a metal rod with a spark gap at its centre that connected two metal spheres. Applying charge to the gap caused high-frequency currents to oscillate while jumping across the gap and along the rods. He detected the waves as damped oscillations via a loop in the far field. What he observed was a form of unenclosed standing wave attached to the wires that eventually ran out of energy. In his analysis he used a mathematical potential now known as Hertz potentials and was thus also able to mathematically investigate the form of the far-field waves. These potentials are the origin of the spinors used within SFT, where a rotation of fixed angular speed ω around a circular path of radius r can be expressed as a complex exponential $re^{j\omega t}$. The use of potentials rather than Maxwell's E and H fields was driven by the widespread motivation to understand EM in terms of the wave equation that featured in applications to fluid dynamics such as water and sound. In the late 17th century Newton had investigated air and water and explained the relationship between the stresses and strains in various fluids, thereby initiating the mathematical and computational study of fluid dynamics.

D'Alembert in 1747 had proposed a wave equation for a vibrating string such as a musical instrument. In the early 19th century Laplace had established *potential theory* in the fields of astronomy and fluid dynamics. Laplace's equation is a generalization of d'Alembert's wave equation that includes $\nabla^2 = \partial^2/\partial^2 x + \partial^2/\partial^2 y + \partial^2/\partial^2 z$, the Cartesian form of the Laplacian operator. He put forward a nebular hypothesis for the formation of the Solar System from a globular mass of incandescent gas rotating around an axis through its centre of mass. The mass cooled and contracted leaving successive rings. These rings also cooled and condensed into the planets, leaving the Sun as the remnant core. Thus Laplace, like Faraday, was another who perhaps saw a unity emerging within physics. One of Laplace's students, Poisson, was equally diverse in his studies. He also studied potential theory due to sources, the non-homogeneous form of Laplace's equation, now known as Poisson's equation, the finite difference method, and differential and partial differential equations. In space-time, the d'Alembertian symbol $\Box$ is equivalent to the Laplace operator within Minkowski space. Some of the wave equations concerning Maxwell's equations are pertinent to the subsequent quantum mathematics that followed. In free space in the absence of sources we get

$$\left(\nabla^2 - \frac{1}{c^2}\frac{\partial^2}{\partial t^2}\right)\mathrm{E} = 0 \tag{1.1}$$

$$\left(\nabla^2 - \frac{1}{c^2}\frac{\partial^2}{\partial t^2}\right)\mathrm{B} = 0 \tag{1.2}$$

These two equations directly demonstrate the wave nature of the EM fields and led Maxwell to his prediction of waves in free space moving with the speed of light. By suitable algebraic and vector manipulation similar wave equations can be derived in terms of A and Φ, the electric vector and magnetic scalar potentials where sources are present.

$$\left(\nabla^2\mathrm{A} - \frac{1}{c^2}\frac{\partial^2\mathrm{A}}{\partial t^2}\right) - \nabla\left(\nabla\cdot\mathrm{A} + \frac{1}{c^2}\frac{\partial\Phi}{\partial t}\right) = -\mu_0\mathrm{J} \tag{1.3}$$

$$\left(\nabla^2\Phi + \frac{\partial}{\partial t}(\nabla\cdot\mathrm{A})\right) = -\frac{\rho_0}{\varepsilon_0} \tag{1.4}$$

These second-order potential wave equations apply to different classes of solution by association of a particular gauge. They are considered as complete as the original first-order Maxwell's equations. In our present time the overall problem remains widely considered to be reduced as the E and H fields have three components that each need solution; on the other hand the A and Φ potentials have a total of only four components. This is a numerical consideration.

Unfortunately for the analytic solutions that still remain latent, the potential equations are second order and not first order, as are the original Maxwell equations. The fact that quantum theory subsequently went down the potential path rather than staying with the original E and H fields of Maxwell's four equations was probably more due to numerical considerations than any other factor. It must be added, however, that the early researchers needed to understand the equations as waves, and this was also a factor. Whatever its origins, this historical choice meant that the second-order potential wave equations with its gauge theory would be associated with the quantum theory of the atom in the 20th century. Given the technological implications, it is no exaggeration to say that 20th-century history may have been very different. The EM knowledge initiated by Coulomb, Ampere, Faraday, and Maxwell was not yet mature.

It is a frequently overlooked fact that the history of quantum theory did not actually start with Planck's analysis of the black box radiator. The electron was first considered a quantum of electricity in 1838 by Faraday who discovered the cathode ray. Various areas of 19th-century mathematical research were crucial in the development of the systematic algorithms that evolved to became quantum mechanics: (1) Matrix methods of solution including the Gaussian elimination method were discovered by Gauss in 1800. (2) Fourier in 1828 noted that discontinuous functions could be written as infinite series of continuous functions. Others including Lagrange, Dirichlet, and Navier helped develop this series technique. The Fourier series was perhaps the first and most important milestone in the long evolution of the quantum theories. (3) Boltzmann and Maxwell invented statistical mechanics for the study of heat in atomic systems. With this a macroscopic system could be related to the average motions of atomic level particles. (4) Hilbert studied the eigenvalues of infinite matrices consisting of series of integral operators and the multidimensional spaces they occupy. In retrospect it can be seen these mathematical methods were perhaps a solution looking for a problem. If the EM fields of atomic outer shell electrons would remain analytically unsolved, the numerical methods developed throughout the 19th-century gave physicists a way to explore the atom including its use as a weapon in the upcoming century.

The ancients, including Aristotle, considered the speed of light to be infinite. Some ancient scholars did consider the speed of light finite but were unable to detect a time of travel. It was Roëmer who detected a delay in the time it took Jupiter's moon, Io, to reemerge from Jupiter's shadow. This he did in summer and winter as Jupiter's distance from Earth varied. From this he calculated the speed of light to be 2.12×10^8 m s^{-1}. In 1879 Michelson, using his interferometer, measured the speed of light as 2.997924560×10^8 m s^{-1}, to an amazing accuracy of 1.1 m s^{-1}. The interferometer split two beams and recombined them via a semi-opaque (half silver) mirror and two reflecting mirrors. The detector discerned interference of the beams to within a half-wavelength and used starlight to detect changes in free space. Michelson and Morley applied the device across the sky to see if a fluid-like aether existed within the cosmos and detected no discernable change in the beams wherever they pointed the device. At the time this seemed to indicate that light waves were unlike sound waves. The experiment also provided evidence that the speed of light did not depend on direction. This has been understood by the overwhelming majority ever since to mean that there was no ether as Earth's motion should have produced a measurable difference in speed. To explain the null result of the Michelson–Morley experiment, Fitzgerald in 1889 suggested that an object might appear foreshortened. Lorentz in 1892 gave an equation for this effect for an object of length L moving at velocity v relative to an ether moving at velocity c, $L' = \sqrt{1 - v^2/c^2}L$. He later added time dilation, and these equations became known as the Lorentz transformation

Given the contemporaneous discovery of radioactivity there was a considerable state of flux in science in the 1890s, when quantum physics was first observed. The photon's discrete behaviour emerged with the failure of science to provide a consistent theory for the energy of a blackbody cavity. Both Wien at short wavelengths and Rayleigh and Jeans at long wavelengths had experimentally obtained differing analytic equations for the energy. The Rayleigh–Jeans equation indicated an "ultraviolet catastrophe", an infinite energy at wavelengths around a micron. Planck resolved the situation by modelling the blackbody's walls, its atoms, as EM dipoles using the potential theory formulated by Hertz along with the concepts of probabilistic thermodynamics developed by Maxwell and Boltzmann. Planck had first acted in an "act of desperation" but slowly came to realize the theoretical implications of the effect that needed to be treated as a series of discrete frequencies rather than a continuously analytic function of frequency. Thus Fourier's mathematics was now supported by blackbody radiation. The Rayleigh–Jeans law for low-frequency radiation intensity emitted by a blackbody is

$$I_{\mathrm{RJ}}(\nu, T) = \frac{8\pi \nu^2}{c^3} kT \tag{1.5}$$

The Wien approximation at short wavelengths is

$$I_{\mathrm{W}}(\nu, T) = \frac{2h\nu^3}{c^2} e^{-h\nu/kT} \tag{1.6}$$

Planck's quantum law is

$$I(\nu, T) = \frac{8\pi \nu^2}{c^3} \frac{h\nu}{e^{\frac{h\nu}{kT}} - 1} \tag{1.7}$$

Planck substituted a series expansion for the exponential

$$e^{\frac{h\nu}{kT}} = 1 + \frac{h\nu}{kT} + \frac{\left(\frac{h\nu}{kT}\right)^2}{2!} + \frac{\left(\frac{h\nu}{kT}\right)^3}{3!} + \cdots \tag{1.8}$$

showing that for low frequencies $h\nu \ll kT$ and high frequencies $h\nu \gg kT$ the classical and quantum laws agreed. Quantum physics and its mathematics had finally shown up on the human radar.

1.3 1903–2010

The years 1903–04 are not usually given as the start of the development of modern physics as we know it today. Yet in retrospect these were the years when Abraham and Lorentz first attempted to apply what we refer to today as SFT. This is the problem of the back-reaction on a particle's own motion. They and later others, including Dirac, chose to study the electron seeking to find a physically plausible model of the electron as a spherical surface of current density that was moving at relativistic speeds. They found that classical electromagnetics resulted in a "4/3 problem" where force on the electron contradicts Newton's law $F = ma$

$$F_e = \frac{4}{3} m_e a - \frac{2}{3} \frac{q^3}{c^3} \dot{a} + \text{"structure terms"} \tag{1.9}$$

To remove this and other unphysical inconsistencies, many tried to modify the classical Lorentz–Abraham–Dirac equation. It appeared at that time that classical electromagnetics had failed. Yaghjian recently revisited the problem and found inconsistencies with *both* quantum theory and classical electromagnetics as the electron radius approaches zero. He found that the generic solution of a charged surface of finite radius moving at relativistic speeds was correct within the macroscopic domain

The underlying self-field problem has remained an unsolved problem until recently, when its modern variant succeeded in solving the hydrogen atom as demonstrated in the next chapter. Yet this open wound on the face of classical electromagnetics did force early 20th-century physicists to think laterally inventing quantum theory, in particular QED. Thus SFT has a pride of place at the start not only of quantum mathematics but of particle physics and the search for exotic particles within the cosmos.

Along with the development of quantum mathematics, Einstein's discovery of the relativity of light, both special (SR) and general (GR), was the theoretical highlight of the 20th century; Einstein played a crucial role in both areas. His 1905 paper on SR assumed that (1) physics is unchanged across inertial frames and that (2) the speed of light c is unchanged across inertial frames regardless of whether the light is emitted by a body at rest or by a body in uniform motion. From these postulates he derived the Lorentz transformations, and the Fitzgerald–Lorentz time and space dilations. Einstein also showed that any ether was superfluous to his deductions. SR was the culmination of efforts to understand why Maxwell's equations were invariant to the Lorentz coordinate transformation. In the same year, 1905, Einstein via the photoelectric effect observed that radiation itself, not just its frequency, acted as discrete quanta or particles of energy $E/h\nu$, termed photons. Bohr, recognizing that the Plank–Einstein equation $E = h\nu$, held for emitted as well as absorbed energy, put forward a quantum theory of atomic spectroscopy in which angular momentum must be whole numbers of Planck's quantum number $\hbar = h/2\pi$.

In the early 1920s Bose used Einstein's photon demonstration of the photo-electric effect and Planck's photon gas of the blackbody law of radiation. Bose developed a statistics for photons different to that of Maxwell–Boltzmann for separated or "distinguishable" atoms. Thus a quantum statistical effect was postulated for photons. A European journal rejected Bose's work as the reviewer could not understand it. In 1924 Bose sent Einstein the paper, enabling publication. In this form of particle statistics the quantum wavefunctions are symmetric. In 1926 Fermi and Dirac introduced a third statistical measure that was applied by Fowler to white dwarfs and by Sommerfeld to electrons in metals. The wavefunctions were in these cases anitisymmetric. The particles appear indistinguishable if the concentration of particles $(N/V) \geq n_q$, where n_q is termed the quantum concentration.

While Maxwell's classical EM equations yielded no solution for even the simplest atom, that of hydrogen, progress in using the derived potential wave equations yielded to partial solution. In 1925 Schrödinger proposed a wave equation relating the kinetic, potential, and total energies of an atomic system to its motions. A computational procedure using a form of Fourier transform pairs was discovered by Heisenberg This method was based on matrix methods that computed the differences between atomic energy states. It soon became clear that these methods represented a huge scientific step forward. As well as experimental methods and theory, computations relating to the physical world could now be obtained; for instance it was possible to obtain chemical bonds within atoms and molecules to within a numerical accuracy. While Schrödinger's *wave mechanics* allowed quantum states to be understood and determined analytically where applicable, Heisenberg's systematic *matrix mechanics* gave a major stimulus towards the computerization of the mathematics. Heisenberg felt that there must be discontinuous quantum jumps to represent the photons in the field yet wave mechanics was based on continuous waves and this led to a division across physics. The impasse ended in 1926 via the name quantum mechanics and Dirac who showed the similarities between the two perspectives. Dirac used a "second quantization" for the EM field with annihilation and creation operators to model the vacuum as an infinite sea of particles that possessed negative energy The resultant theory was compatible with special relativity and introduced spin to quantum mechanics. Yet how to physically interpret a wave theory combined with quantization of both particles and field has remained enigmatic. De Broglie in 1923 showed that electrons as well as photons in fact all matter had a wavelike nature he termed matter-waves. This was a mathematical statement of the particle–wave duality that was strongly supported by experiment.

Looking at the method of QM, take the example of the N-particle Schrödinger equation

$$i\hbar\partial_t\Psi(r_1, r_2, \ldots, r_N, t) = \frac{\hbar c}{i}\Psi(r_1, r_2, \ldots, r_N, t) \qquad (1.10)$$

where the total Hamiltonian may be written

$$H = \sum_i H(r_i) + 1/2\sum_{i\neq j} V(r_i - r_j) \qquad (1.11)$$

and where $H(r_i)$, the self-Hamiltonian for each particle, may be written in terms of its self- kinetic and potential energy:

$$H(r_i) = -\frac{1}{2m}\nabla^2\Psi(r_i, t) + V(r_i)\Psi(r_i, t) \qquad (1.12)$$

While Schrödinger, Heisenberg and Dirac all used a first quantization for the particles, various fields including the EM field were left in their classical form. In 1927 Dirac quantized the vector potential from which the radiation field in a Coulomb gauge could be obtained.

$$A = \sum_\lambda (q_\lambda A_\lambda + q_\lambda^* A_\lambda^*) \qquad (1.13)$$

This process of taking Fourier components of a field, for instance a series of plane waves, is termed second quantization. The history of QFT began with the evolution of Schrödinger's non-relativistic equation to Dirac's relativistic equation. Schrödinger expressed the phase of a plane wave as a complex factor (1.14). The Klein–Gordon equation (1.15) is a relativistic version of Schrödinger's equation. Using Feynman notation (1.16) we see how the Klein–Gordon equation can be analytically factorized to yield the positive and negative energy versions of the Dirac equation (1.17), where the γ_μ are 4×4 bi-spinors. Dirac thus predicted the electron's anti-particle, the positron, via these equations. Although Dirac's equation is first order it is derived from a second-order wave equation, and it is this that demands gauge symmetry and invariance under conformal transformations.

$$i\hbar\frac{\partial}{\partial t}(x, t) = \widehat{\mathrm{H}}\Psi(x, t) \qquad (1.14)$$

$$\left[\nabla^2 + \frac{m^2c^2}{\hbar^2}\right]\psi(r) = 0 \tag{1.15}$$

$$(\partial^2 + m^2)\psi = 0 \rightarrow (i\partial + m)(i\partial - m)\psi = 0 \tag{1.16}$$

$$(\gamma_\mu\partial_\mu + im)(\gamma_\mu\partial_\mu - im)\psi = 0 \tag{1.17}$$

Second quantization can be applied to the ψ functions of the Dirac equation (1.17) and is used in many-electron and many-atom problems. The mathematics continues by defining a Lagrangian based on conjugate pairs from which "observables" are obtained. Thus relativistic QM became *the* tool for mathematical physicists. The *hole* theory invented by Dirac grew out of the difficulty that a free electron could assume any energy state from $-mc^2$ to $-\infty$. Two assumptions were made to avoid this outcome:

(1) All negative energy states from $-mc^2$ to $-\infty$ are filled up with *other* electrons. No electron can jump into one of these states; electrons satisfy the Pauli exclusion principle: *In a closed system, no two electrons can occupy the same state.* This principle is usually applied to atoms containing many electrons, not to an "infinite sea" of particles. In such atoms a similar QM complication arises, in that an electron n orbiting about a proton in the Bohr model, there can be $2n^2$ distinct states that can be occupied.
(2) The electrons filling up these negative energy states do *not* produce any external field or contribute to the charge, energy, or momentum of the system.

Hole theory led to the discovery of the positron (see (1.17)) and had other far-reaching consequences, including the creation and annihilation of particles in an energy-rich vacuum in which particles drift in and out of existence. Pair production was a verification of the physics predicted by QED. Zero-point energy is an effect upon the radiation field itself. Both hole theory and zero-point energy involve subtracting energies from the overall Lagrangian. The concept arises of eliminating *virtual* particles that exist temporarily. Thus the discovery of QED and QCD occupied theoretical physicists till the end of the century. Various gauge theories were proposed to apply as fields in nuclear and atomic regions.The application of the statistical methods of QFT to the Standard Model of particle physics was a major achievement for late 20th-century physics.

Perhaps the most influential physicist of the pre- and early quantum era apart from Einstein was Heisenberg, who with Bohr established the basis for the mathematics of QM. In 1927 Heisenberg stated his uncertainty principle (HUP) that in the process of measuring the position x of a particle, its momentum p is disturbed so that $\Delta x \, \Delta p = h/2$. According to Heisenberg it is impossible to know anything about the photon. Thus Heisenberg and Born stated the Copenhagen interpretation

> We regard quantum mechanics as a complete theory for which the fundamental physical and mathematical hypotheses are no longer susceptible of modification.

—Paper delivered to Solvay Congress of 1927

Einstein in 1936 replied, "I still do not believe that the statistical method of the Quantum Theory is the last word, but for the time being I am alone in my opinion."

Heisenberg wrote in 1969:

> To those of us who participated in the development of atomic theory, the five years following the Solvay Conference in Brussels in 1927 looked so wonderful that we often spoke of them as the golden age of atomic physics. The great obstacles that had occupied all our efforts in the preceding years had been cleared out of the way, the gate to an entirely new field, the quantum mechanics of the atomic shells stood wide open, and fresh fruits seemed ready for the picking.

In the 1920s to '30s the mathematics of quantum theory emerged in the full glare of the scientific spotlight and was correctly seen as revolutionary. The numerical methods such as those invented by Slater and Hartree–Fock that evolved to solve the equations associated with QM gave motivation for the invention of the computer and the realization of atomic energy. As discussed by Heitler and others, major limitations to which QFTs were prone were soon discovered, such as inaccuracies, infinities, and huge interaction matrices. "Renormalization" is a numerical procedure that essentially eliminates singular integrals from the computations of QED. Viewed as a numerical "fix" by its originators, renormalization is today seen as a way to apply "self-consistency". Self-consistency is as much a field of logics as a numerical technique. In fact renormalization is closely related to the attempts by Abraham, Lorentz, and Dirac to find a consistent point limit for the electron within classical EM theory and QFT. There is a formalism contained within QFT that provides a considerable degree of elaboration in the mathematics that seeks to model a physical application.

Another question of considerable debate that arose in the early days of QFT is the mass of the photon. In gauge-invariant electrodynamics the photon is assumed to be massless since, simply, the mathematics of the field theory is considered to be "in trouble" otherwise. Some modern theoretical physicists have even tried to mandate the photon to be standardized as massless, perhaps in part to avoid the problems associated with the existence of a massive photon. Jackson, in the third edition of his text on classical electrodynamics, reviewed the search for an experimental upper limit for the photon mass. Such an upper limit is considered the cut-off between classical and quantum electrodynamics. There is recently a groundswell of opinion to support a massive photon, including Lehnert and Roy, who in 1998 proposed an extension to conventional EM theory based on the concept of space charge in vacuo.

From the prism of quantum theory many of the ideas in the above paragraphs might appear complete and without need of further comment or refinement. Their classical roots within 18th-century field theory, 19th-century mathematics, and the early days of blackbody radiation have been long forgotten. But problems are still there, imbedded inside current quantum mathematical methods.

While it is not the aim to provide complete discussion in this introductory chapter, it is profitable to briefly outline the perspective from within SFT concerning the issues mentioned above. Essentially, SFT and its concepts go "beyond quantum" and lead to numerical improvements and analytic methods where there are currently none. By proposing an internal hydrogenic structure for the photon, SFT sees beyond the numerical inaccuracies associated with HUP and can thus clarify the physical processes as presently envisaged via QFT. The bi-spinorial centre-of-motion fields used by SFT obviate the entire issue of renormalization as

EM particles never reside at their origin but rotate around a centre of motion. Thus EM fields came as pairs of motions and apply to pairs of particles that have an effect upon each other. The concept of the EM self-field can be applied to update QM and the QFTs, so that actual orbits and not just probabilistic orbital densities can be obtained. In brief, SFT clarifies both the theoretical and the numerical solutions due to current implementations of QFTs. Moreover, concepts like Dirac's hole theory and the Pauli exclusion principle have corresponding mathematical concepts within SFT. One pertinent concept in SFT is "conjugate" spin, whereby a particle such as an electron or a proton in the hydrogen atom is associated with two photons that bind the two particles together. One of these photons has a spin, the conjugate of the other. Collisions between the photons and particles flip the sign of the photon spin. Hence, since there is a phase difference of 90° between the photons, the overall Lagrangian will be unaffected if the photon turns out to have a tiny but significant mass. Hence QED is not troubled by a massive photon since the energies of the two photons cancel each other out and the gauge invariance is maintained; similar concepts relate to symmetry, which is conserved. Another mathematical outcome of SFT is that particles of charge of like sign can also exist as stable pairs given the appropriate energy density; this is a basis for Pauli's exclusion principle. Inside nuclei these SFT spin concepts apply to the weak nuclear forces and allow for stability of three particles within the strong nuclear regions.

While Einstein's SR and GR were revolutionary steps forward in understanding light and the cosmos they too have their difficulties in their present form. Physicists might accept the constancy of light across an isotropic solar system as a near-field phenomenon supported by experimental evidence. However, others might reject both the constancy of light and the isotropic constituency of the universe across the entire universe at the present epoch. While the cosmic background radiation can show the early universe to be isotropic to 1 part in 100,000, it may be expanding in a flat but anisotropic ellipsoid that varies across its geometry especially towards its edges. Similarly some physicists understand the warping of space-time near gravitational masses as a region where conundrums and paradoxes can occur such as the *twin paradox*. This leads to the possibility of time travel via "wormholes" through the fabric of space-time joining two remote locations of gravitational intensity *black holes* existing near the centres of galaxies.

With regard to relativity, the bi-spinors of SFT apply equally to the photon as to the EM field. Thus SFT sees relativity with its time and distance dilations as a physically straightforward outcome resulting from the internal motions of the photon. In SFT there is no twin paradox per se. The dilation is an effect that causes light to change spin and radial mode and is internal to the photon due to the energy density in a region impinging on the photon. In terms of time, *absolute time* does not change from point to point in the universe, but relative time due to the rotation of the photon does vary. The twins who meet somewhere in space-time should be the same absolute time unless of course there is a discernable *biotemporal* reaction of the body to the varying speed of light from point to point across the universe. When the twins reunite they will be the same relative age once again, assuming the biotemporal effects of space-time travel are completely reversible.

Redshift is a phenomenon that applies to EM waves that lose energy by several distinct means such as refraction, reflection, and absorption. Since each is a shift towards the longer wavelengths, they are called Doppler effects, similar to the frequency lengthening of a sound moving away from the hearer. It is considered that the farther away a light source is, the larger the redshift. Quasars are very powerful sources of EM radiation, and these appear to be very distant, as inferred from their measured redshift. This is considered to be due to the expansion of the universe assumed geometrically to be similar to a flat disc, like our Solar System and our galaxy. The cosmic microwave background radiation has a redshift of more than 1000 corresponding to a distance of more than 13 billion light years. SFT with its non-negligible photon mass suggests some interesting mechanisms that do not change the prevailing Big Bang theory of the origins of the universe.

Newton and Einstein remain the most important figures in the search for a mathematical description of gravitation. Newton's inverse square law is based on Pythagorean distances and applies to terrestrial effects such as the apple falling from the tree. According to Einstein's relativity, the distances and times that can be dilated are considered externally measurable. However, in SFT the photon performs both internal and external motions that can be modeled by bi-spinorial rotations. The internal motion of the photon is thus hidden at macroscopic levels. This insight allows us to understand relativity in a physically plausible and intuitively acceptable manner. Free *empty* space is *not* warped, only in the eye and mind of the observer who sees these relativistic motions as warped. Seeing is not believing at relativistic speeds. Warping is an effect entirely within the retina and mind of the observer due to the way the photon rotates internally and externally. The gravitational fields are indeed different in a stronger gravitational field, but empty space is unchanged; only the space-time is warped. This may seem a subtle difference, but it is a statement about the way mathematical coordinate systems transform, about the role of photons in that process, and about the space in which the physics takes place. Unlike quantum theories that see zero-point vacuum space as a medium where virtual particles flash in and out of existence, SFT separates these short-lived particles and anti-particles out from the background empty space. Space is space wherever it is.

More fundamentally, SFT sees gravity as a differential effect due to the various forms of matter. According to SFT various forms of matter are involved within different levels across the universe. At the smallest level of gravitation, the level of solar systems, EM field is balanced in a differential sense. At the next level up according to SFT, the level of galaxies, the strong nuclear field is likewise balanced. At the second level up, the level of the cluster galaxies, another as yet unidentified field is balanced. There is reason to believe that at the level of the universe, universes are balanced by another unknown field. According to SFT there is a tiered gravitational structure involving various forms of matter and the types of fields or bosons with which they interact. This tiered structure is responsible for the different forms of gravitational motion seen within the cosmos. This may be the cause of the tiny inhomogeneity seen within the plot of cosmic microwave background (Fig. 1.10).

There is another issue known as the EPR paradox that is at the heart of the way SFT sees physics. Einstein challenged quantum theory, which he saw as incomplete, in a paper that contained the paradox. As a consequence of this historical debate, quantum

Figure 1.10. Cosmic microwave background temperature fluctuations over the full sky. (Credit: Wikimedia.) See also Color Insert.

entanglement is now understood as a way in which parts of a QM system are connected. The quantum states of the constituent parts are linked together, so that one part cannot be properly described without mention of all other parts within the system. Einstein referred to this as "spooky action at a distance". SFT too sees quantum theory as incomplete. As we have seen, SFT suggests a missing coordinate within the photon, which is the underlying reason for HUP and its lack of knowledge concerning photons. Within atomic and molecular physics according to SFT, photons link all atoms together as in a crystal lattice. This is the underlying photon level of interaction applying to many, if not all, atoms within the universe. This applies to crystals where all atoms rotate in synchronized fashion and to biological systems where photons link to other objects in the view of the biological system. This usually entails the biological capacity for optical sight but can encompass other forms of biophotonic emission and detection. This includes the complete body-fields of biological systems in general but is found in specialized physiological tissues such as the well-known elasmobranch fish that use pores on the lower peripheries of their fins for predation. It may also include tissues within the cranial tissues of various species, including humans. Universal "consciousness" is related to this photonic physics.

Perhaps *the* intrinsic difference between earlier physical theories, including the various quantum theories dominant during

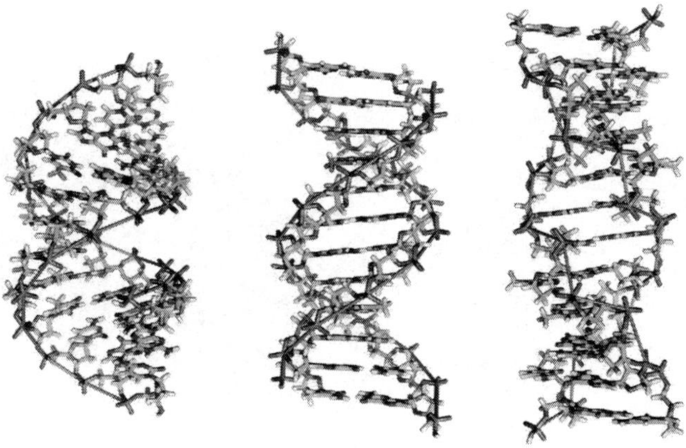

Figure 1.11. Piecewise bi-spinorial structure of the double-helix of the DNA. (Credit: Wikimedia.) See also Color Insert.

20th-century science, and SFT is its application to biophysics. Microbiology reveals the genetics of life centres around deoxyribonucleic acid (DNA) molecule. It was an achievement of the 20th-century biological science that DNA was determined via X-ray diffraction to be a double helix. Empirically SFT provides precise piecewise bi-spinorial field motions that can yield such doubled stranded helices within molecular binding structures (Fig. 1.11). SFT fields are described as "beads-on-a-string"; this is also the terminology used to describe histones stored within the packed DNA (Fig. 1.12).

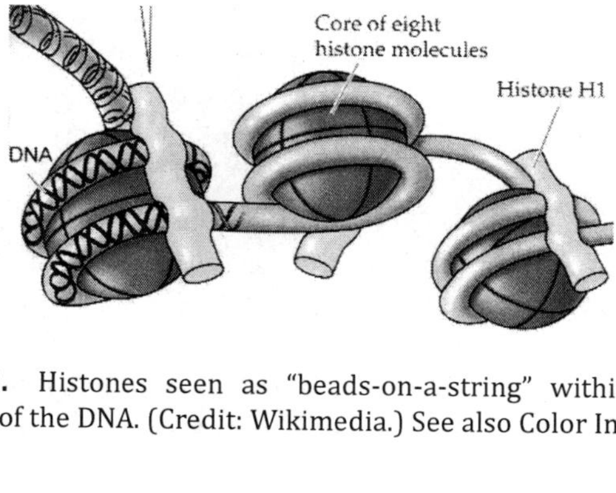

Figure 1.12. Histones seen as "beads-on-a-string" within the coiled double helix of the DNA. (Credit: Wikimedia.) See also Color Insert.

Chapter 2

Self-Field Theory

During 1903–04 the effect on the motion of a single particle due to its self-field was studied by both Abraham and Lorentz. They modelled the *isolated* electron as a charged spherical surface of finite radius and found inconsistencies with classical electromagnetic theory as the radius approached zero. In 2005 the self-fields of *pairs* of particles were finally understood as a mutual phenomenon. The singularity problem at the charge points was solved by using motions that avoid the charge points, assuming that at equilibrium the two particles rotate never residing at their own centres of rotation. The self-field theory model[1] for the electromagnetic field as a cyclic stream of photons provides an analysis of the hydrogen atom and yields a derivation of Planck's number $\hbar$. The bi-spinorial function for each particle provides a physically plausible interpretation of relativity. The "beads on a string" stream-like electromagnetic field modifies the macroscopic time-invariant field laws of Coulomb and Ampere at the atomic level. The E- and H-fields must be measured between centres of rotation rather

[1]Appendix A details the mathematical preliminaries including methods of solution, spinors, the centre-of-motion (COM) E- and H-fields, imaginary numbers within 3- and 4-D coordinate systems, and the lack of errors in SFT due to uncertainty, seen in SFT as a particular form of numerical modelling error associated with an over-constrained model of the photon.

Self-Field Theory: A New Mathematical Description of Physics
By Tony Fleming

www.panstanford.com

than between charge points and applied as a coupled complete electromagnetic field. The atomic self-field motions are obtained using the Maxwell–Lorentz (ML) equations. Quantum theory can be reinterpreted to include the coupled bi-spinorial field to yield the same deterministic closed form eigensolutions as self-field theory. Space–time orthogonality shows the complete self-field theory outer shell electronic structure to be analytic. Self-field theory allows reinterpretation of the weak and strong nuclear forces via a modified system of ML equations.

2.1 Introduction

Present-day understanding of the hydrogen atom is linked to the quantum theory that evolved during several decades of effort from the late 19th century until 1927. Bohr put forward a quantum theory of spectroscopy in which angular momentum is a whole number multiplied by Planck's quantum number $\hbar = h/2\pi$.[2] Using Bohr's theory, spectral lines can be expressed as a quantum series, for example, the Balmer series,

$$\nu_{mn} = R\left(\frac{1}{m^2} - \frac{1}{n^2}\right)$$

where $m^2 = 4, n = 3, 4, 5, \ldots$ and Rydberg's number $R = \frac{q_e^4 m_e}{8\varepsilon_0^2 h^3 c}$. With this theory, the ground state energy of the hydrogen atom, Bohr's energy, $E_B = Rhc$, and the electron's expected position, the Bohr radius, $r_B = \frac{\varepsilon_0 h^2}{\pi m_e q^2}$ can be estimated. Spectroscopic experiments led to an understanding of how the hydrogen atom was excited by E- and H-fields.

Despite intensive investigation this same period saw a complete failure to find any way in which atomic physics could be based on classical electromagnetic (EM) theory. Maxwell had previously reached the conclusion that gravitation could not be based on EM theory despite the similarity between the inverse-square forms of Coulomb's electrostatic force and Newton's gravitational force. Concepts of planetary motion led only to unstable spiraling of the electron into the nucleus at the centres of atoms. The alternative is that the atom slowly spirals apart into oblivion. Cosmological

[2] Appendix B lists relevant physical constants, variables, and values.

concepts of the Big Crunch or the Big Freeze are similar in nature. A conviction against EM as a basis for atomic physics grew as the mathematical methods of quantum mechanics (QM) evolved.

By 1927, QM theory had been developed as the eigenvalue equation $\hat{H}\psi\,(r,t) = j\hbar\frac{\partial\psi}{\partial t}$, where $\hat{H}$ is a Hamiltonian associated with a particular application. QM was based on the probabilities of large numbers the statistical mechanics of Maxwell–Boltzmann, the variables were expectations. QM helped detect four quantum numbers n, l, m, and s associated with the electron. From the late 1920s to the mid-1940s quantum electrodynamics (QED) evolved as a hybrid of EM, quantum theory, and special relatively (SR). Due to the uncertainty principle, quantum field theories (QFTs) including QED cannot be solved deterministically. At their core are quadratic functions based on quantized EM potential solutions to the EM wave equations. In almost all cases QM requires numerical solution methods. As well as the second-order partial differential equations, boundary conditions at infinity and near singularities need formulating. Applications require evaluation of significant terms among infinite numbers of integrals using Feynman diagrams and renormalization. Quantum chromodynamics (QCD) in turn is a generalization of QED devised to handle the added particles falling within the standard model of particle physics observed in the last half of the 20th century. Quantum numbers such as “flavour” and “colour” exist in addition to those due to electronic interactions with the nucleus. Despite enormous numerical difficulties, the standard model of the hydrogen atom resulting from the Schrödinger equation has until recently been the only option. Certainly, no EM solution based on the dynamics of charged particles has been useful in modelling the hydrogen atom apart from Bohr’s early model. Yet classical EM theory remains intrinsically at the heart of current QFTs.

In 1954 Einstein wrote of classical EM theory and quantum theory:

> In the year nineteen hundred, in the course of purely theoretical (mathematical) investigation, *Max Planck* made a very remarkable discovery: the law of radiation of bodies as a function of temperature could not be derived solely from the Laws of Maxwellian electrodynamics. To arrive at results consistent with the relevant

> experiments, radiation of a given frequency f had to be treated as though it consisted of energy atoms (photons) of the individual energy hf, where h is *Planck's* universal constant. This discovery became the basis of all 20th-century research in physics and has almost entirely conditioned its development ever since. Without this discovery it would not have been possible to establish a workable theory of molecules and atoms and the energy processes that govern their transformations. Moreover, it has shattered the whole framework of classical mechanics and electrodynamics and set science a fresh task: that of finding a new conceptual basis for all physics. Despite remarkable partial gains, the problem is still far from a satisfactory solution.

In hindsight it is easily seen how quantum phenomena occur in nature. The eigenvalue problem had been examined by Cauchy, Lagrange, and others in the first half of the 19th century, so discrete solutions were known to physicists. Take the example of two children of different masses balanced on a fulcrum. There is a uniform array of eigensolutions along its length depending on the ratio of the masses. For equal masses eigensolutions exist all along the fulcrum. In the atom the addition effect of two orthogonal forces demands a discrete physics. Nature has many examples of such balances including the atom and the photon; our Universe itself may be another such balance.

Self-field theory (SFT) presents a clear and different perspective to both classical EM and quantum theory (see Figures 2.1-2). Hidden like a nugget of gold in the earth is an exact dynamic solution to Maxwell's equations but only when the correct conditions are implemented, in this case the Lorentz forces and conditions of periodicity. SFT yields analytic closed-form solutions for the electron's motion in the hydrogen atom as well as Rydberg's number, Planck's number, and the Balmer formula. The fields are unconventionally measured relative to centres of motion rather than directly between charge points as in the Liénard–Wiechart potentials. Hence particle motions are defined by two distances and not one as in classical electromagnetics. Although there are several, this is perhaps *the* essential difference between present and previous attempts to solve for atomic motions using electromagnetics. After testing

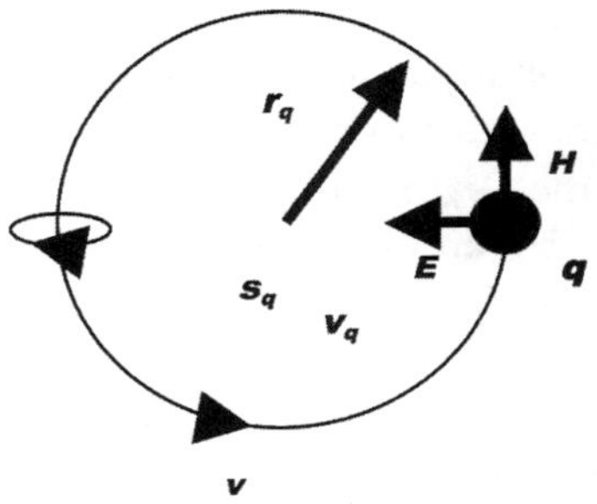

Figure 2.1. A charged atomic particle rotating with respect to the centres of its motions due to the E- and H-fields of a mutual charged particle. Three geometric parameters are associated with the motion, the radius r_q, surface area s_q, and volume v_q of the charge as it performs its cyclotron motions. Illustrated is its orbital and cyclotron motion.

particular bi-spinorial field forms directly via Maxwell's equations, substitution of the bi-spinors into the ML equations results in a general system of algebraic equations in the spinor variables. For the hydrogen atom this introduces a previously undiscovered level of photonics, and simplifies the mathematics used to obtain the physics.

Another essential difference between SFT and QFT can be understood via the analogy between finite difference method (FDM) and finite element method (FEM) well-known to numerical analysts. In numerical solutions one method of approximation can be obtained by the use of the original partial differential equations along with a trial function or series of trial functions for the fields including a Dirac-delta impulse function. It is often a relatively

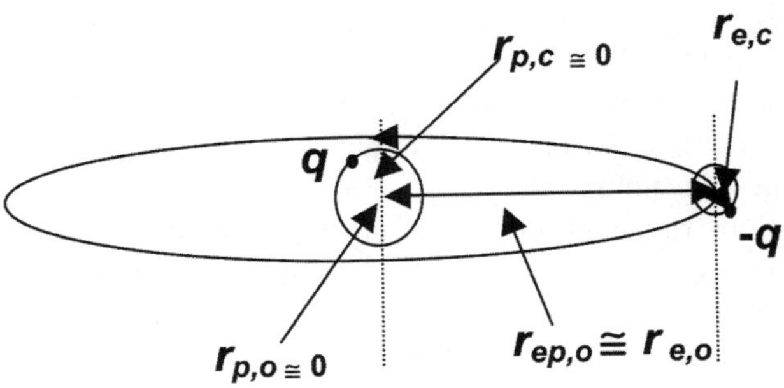

Figure 2.2. Hydrogen atom modelled with an "infinite mass" proton.

simple matter to program these FDMs on a computer. An alternative but more intricate method is to use FEM were the original partial differential equations are integrated over a series of finite elements similar in part to the Feynman diagrams of QFT. By comparison FEM is more difficult than FDM to program but often results in a more accurate representation of the physics. Hence in respect to the analogy between SFT and QFT we use a form of finite differences to the analytic solution in SFT, while an integral equation method is used within QFT. However, as current QFTs do not incorporate a bi-spinorial model for the E- and H-fields they are intrinsically inaccurate and this can be associated with Heisenberg's uncertainty principle (HUP).

Often, with justification, the QED solution for the Landé g-factor is quoted as the most accurate theory ever, with the solution given to many places of significance. Since the solution sought in the case of the electron's g-factor is an area, the truncation error for the QED method turns out to be of quadratic order rather than first order as in other cases where atomic bond lengths might for example be sought. The accepted value for electron spin g-factor is $g_e = -2.0023193043622$ with an uncertainty 0.0000000000015. It should be understood that this value and its uncertainty have been determined using both theory and experimental measurement, a truly amazing degree of validation of both theory and experiment. Nevertheless, the fact is that QFTs do not use a bi-spinorial field function, and thus are inaccurate in their current implementations. It appears, to the author, that there is much scope for incorporating a bi-spinorial field within QFTs to perhaps obtain more accurate models of physics than the current version of SFT.

While the mathematical eigenvalue formulation of SFT is classical its model of electromagnetics is certainly not. In part, this is reminiscent of the early days of quantum theory when Planck first wondered if the EM frequencies might be discrete within blackbody radiation. The fields in SFT are discrete field particles, photons, rather than the continuous vector fields of Maxwell's classical electromagnetics. QFT also models the fields as discrete quanta. However, each individual field[3] in QFT may be taken over the entire

[3] As discussed in Appendix A the "fields" are actually potentials in QFT, compared with the coupled E- and H-fields in SFT.

range of solid angles connecting any number of particles depending on the geometry and the number of interacting atomic particles. The field is shown as a small wavy line within Feynman diagrams but the QFT mathematics does not specify any actual path, only the start and end points where Dirac-delta functions are used to insert propagator kernels and Green's functions depending on the field and its geometry. Basically the fields are modelled as impulse functions specified at charge points; the field is invariant over solid angle even though the Feynman diagram indicates a beam-like wavy line. Within SFT, a pair of particles defines the bi-spinorial field and its motion, this pair and its bi-spinorial field form a unique couple. The fields of this couple do not influence any other charges apart from the charge couple. The transit of the SFT field between charges is specified via the bi-spinorial function. It assumes various motions including spiral-helices as it transits between the electron and proton within the hydrogen atom. Certainly there is a vast difference between this SFT space–time variant field motion (see Figure 2.3) and the time-invariant classical electromagnetics where the field ubiquitously covers all solid angles with no definition other than its vector nature as to the actual field motion, field flux being the only indicator of field motion. Similarly, the uncertainty of the field within QFT is related to its lack of a complete and coupled EM bi-spinorial field form.

There are other major differences including an absence of HUP within SFT. As the photon is modelled via a bi-spinor uncertainty is obviated. As discussed in Appendix A, uncertainty can be considered a theoretical error, a lack of a spatial degree of freedom, a form of

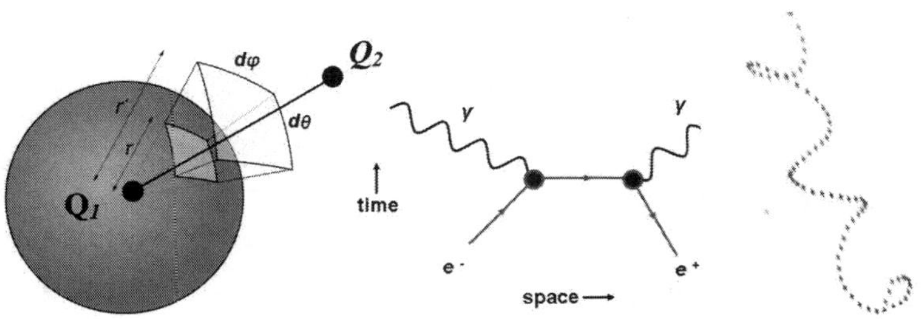

Figure 2.3. Field forms: (a) classical electromagnetic, (b) QFT, and (c) SFT. (Credit: Wikimedia.) See also Color Insert.

truncation error, rather than a geometric modelling error. In SFT the electron's self-fields are modelled via a complete EM function that explicitly includes both E- and H-fields, enabling the complete analysis of the mutual effect of the self-fields of two particles. Unlike the quantum potentials that are expectations yielding probabilistic solutions, the bi-spinorial field variables of SFT allow completely deterministic solutions. This results in a clearer picture of the physics that includes the particle–photon interactions that contain information about the binding mechanism. Based on the first-order ML equations where bi-spinorial fields apply, neither SR nor gauge symmetry is problematic as both are implicit. The bi-spinor field variables of SFT are *a priori* relativistic and its solutions analytic rather than numerical. In EM applications gauge symmetry is maintained by analysing pairs of particles and not just single particles. Similarly a pair of conjugate photons of finite mass do not constitute *symmetry breaking* of the Lagrangian as in QFT. It is worth noting that particle physics concerns deeply inelastic collision dynamics that overcome binding energies, seeking cross-sections, momenta, and energies of the particles while SFT concerns series of elastic collisions inducing dynamic equilibria and periodic motions that allow binding energies to be maintained.

In SFT the speed of light is not constrained other than the experimentally observed speed in free space. Thus, perhaps not surprisingly, Planck's "constant" is found not to be constant but to vary within the SFT theory of the hydrogen atom. Thus an analytic derivation of Planck's number is found by the SFT formulation. Dirac's eigenvalue equation $\hat{H}\psi(r,t) = j\hbar\frac{\partial\psi}{\partial t}$ assumed empirically the postulates of QM including the heuristically determined value of $\hbar$. Apart from its closed-form deterministic solutions for the hydrogen atom SFT supplies an analytic expression for $\hbar$, the number at the core of the theory on which quantum theory rests.

2.2 Self-Field Theory

A mathematical description for the self-fields of charged particles was first derived by Abraham in 1903 and Lorentz in 1904. This is the effect on a single moving charge of any back reaction due to its own radiating field. In electromagnetic self-field theory (EMSFT) *two* charged particles and their self-fields form a closed atomic system; the particles and their mutual fields forming a balanced atomic self-system. This is not an infinite, but a spatially finite, closed system where the photons are contained inside the atom. The standing waves surrounding a cross-dipole representing the hydrogen atom can be shown to have no far-field radiation at certain frequencies and dimensions. A condition of balance is found which is associated with the radiation fields of the cross-dipole such that no Bessel functions of kind 2 are involved; the internal standing waves are self-fields of finite extent rather than far fields of infinite extent. These fields have the same spinorial form as the EMSFT centre-of-motion (COM) fields described in Appendix A.

In general, both the particles and the EM fields that control the motions of charged particles satisfy the ML equations. In atomic physics, regions where particle–field interactions occur are assumed isotropic and homogeneous, and the constitutive parameters, ε_0 and μ_0, the permittivity and permeability of free space respectively, are scalars. Where discrete particles carrying units of elementary charge q of opposite polarity are studied, in the absence of nebular regions of charge and current density, the ML equations are written as

$$\nabla \bullet \vec{E} = \frac{q}{v_q} \tag{2.1a}$$

$$\nabla \bullet \vec{H} = 0 \tag{2.1b}$$

$$\nabla \times \vec{E} + \mu_0 \frac{\partial \vec{H}}{dt} = 0 \tag{2.1c}$$

$$\nabla \times \vec{H} - \varepsilon_0 \frac{\partial \vec{E}}{dt} = \frac{\pi}{s_q} q\vec{v} \tag{2.1d}$$

The Lorentz equation for the field forces acting on the particles is written as

$$\vec{F} = q\vec{E} + q\vec{v} \times \vec{B} \tag{2.1e}$$

The constitutive equations in free space are

$$\vec{B} = \mu_0 \vec{H} \tag{2.1f}$$

$$\vec{D} = \varepsilon_0 \vec{E} \tag{2.1g}$$

The relationship between the speed of light[4] and the ratio of the fields

$$c = (\varepsilon_0 \mu_0)^{-1/2} \tag{2.1h}$$

The atomic energy density per volume is

$$dU = \rho dV = \frac{1}{2}(\varepsilon_0 \tilde{E} \bullet \tilde{E} + \mu_0 \tilde{H} \bullet \tilde{H})dV \tag{2.1i}$$

which depends upon the E- and H-fields in the atomic region. Equations (2.1a–d) are termed the EM field equations.[5] In these equations, v is the particle velocity, m is its mass. It is assumed that the volume of integration v_n over which the charge density is evaluated, and the area the charge circulates normal to its motion S_n, are calculated during successive periods over which the internal motions of the atom take place (Fig. 2.2). At this stage, the periodic motions are not assumed circular, regular, or constant. The area parameters need retrospective examination to check any solutions that are uncovered for a particle's motion. In distinction to quantum theory, the charge- and mass-points are non-singular, as a particle does not to reside at the charge-point due to its assumed bi-spinorial motions.

As well as the motions of the atomic particles, Maxwell's equations specify the spatial distribution of photons that comprise the E- and H-fields due to the presence of charged particles. The concept of the field and the charged particles within it is that of a collection of smaller particles (photons) that transit between larger particles (electrons and protons). We may think of the visual capability of computers to "zoom" into a displayed region. Afar, the field looks like a nebulous cloud. As we zoom in, the cloud comprising the field becomes a series of discrete point-like particles, the photons. At the same time the particles, the electron and the proton, may change their visual character from point charges to

[4] In EMSFT, the speed of light is not proscribed from being variable. Depending on the energy density of the region under investigation, and the photon state, c can vary.

[5] Where a nebular current density is used in (2.1d), the factor 4π comes about from an application of Green's theorem leading to a surface over the volume enclosed by the charge density. For the case of discrete charges, the factor π represents the area enclosed by the moving charge point.

objects with internal dynamic structure.[6] The E- and H-fields acting on the electron for instance are a photon exchange process between proton and electron. The rotating motions are related to this photon exchange. In QM terminology, spinors represent an internal quality of electrons that possess spin. In EMSFT, the electron's spinorial motion results from the mathematics and involves the entire atom acting in a mutual fashion. Spin is therefore an outcome of the dynamic motion due to the H-field. The orbital/cyclotron motions are visualized as discrete photon/electron interactions that change the direction of the electron. The fields are stream-like and exist only between particles.

2.3 Solving Maxwell's Equations Using Bi-spinors

The full system of ML equations (2.1a–i) can be solved exactly to yield analytically cyclic solutions where atomic particles and their E- and H-fields perform rotations in two orthogonal but coupled directions (Fig. 2.4). Because this field pair induces another pair of fields this second pair can, depending on their speed and frequency, in turn produce the original fields. These field pairs are produced by atomic particles each performing two orthogonal motions. The overall result is a "self-perpetuating" field. This approach is general in its application; and can be used as the basis of a photon chemistry where the photon, assumed to have internal structure, has an analytic "Balmer-like" spectroscopic nature, which is continuous rather than discrete as with the spectroscopy of the atom. We now turn to the details of this analytic self-field solution for the hydrogen atom.

Separation of E- and H-field variables or potential variables into azimuthal modes is well known to EM theory using isolated or coupled modes (see Figure 2.4). In the following the principal azimuthal mode ($m = 1$) is investigated. The complete spatial information is retained (2.2) rather than substituting any parametrically related rotation times as in equation (2.6). Following Von Hippel, we assume a separated form for the electron's E-field as a radial spinor rotating

[6]Photons cluster in regions of enhanced energy and can partake in the formation of high-energy nuclear particles.

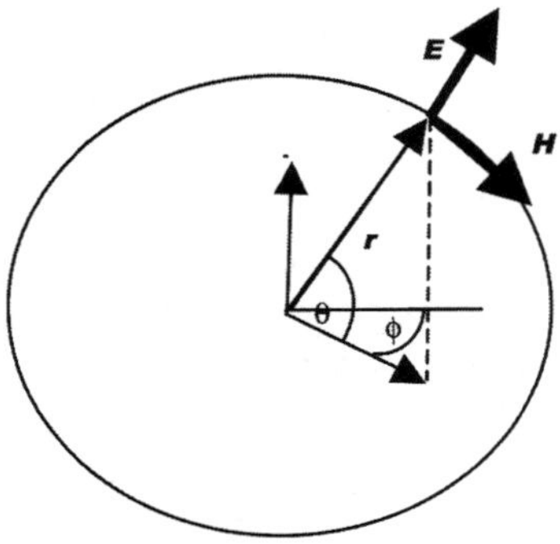

Figure 2.4. Rotating pair of orthogonal E- and H-fields in spherical coordinates.

in the azimuthal angle ϕ:

$$E = \frac{1}{4\pi\varepsilon_0}\frac{q}{r^2}e^{j\phi}\hat{r} \tag{2.2}$$

The E-field can be envisaged as a stream of individual photons all following each other forming an expanding helical spiral as it moves from proton to electron. The stream collides elastically with the electron which changes the direction of the stream. The stream now forms a diminishing helical spiral as the photons move from electron back to proton.[7] Each photon in the stream emanating from the proton tracks the electron as it moves in a double spinor motion; thus, the photon motion is compatible with that of the electron. Equation (2.2) is a dynamic field in contrast to the time-invariant or electrostatic Coulomb field. It is a solution to Maxwell's first divergence equation (2.1a).[8]

The time derivative of the H field can be formed from the curl of the E-field (2.1c). A linear relationship exists between time and angular rotation $\phi = \omega t$. Integrating over time yields an H-field rotating with the E-field.

$$H = \frac{1}{4\pi\varepsilon_0\mu_0}\frac{q}{\sin\theta\omega r^3}ej^{\phi}\hat{\phi} \tag{2.3}$$

[7]The energy density due to photon density is higher near the proton.
[8]The scalar potential φ is related to the E-field ($\varphi = \frac{-1}{4\pi\varepsilon_0}\frac{q}{r}e^{j\phi}$).

This form is a solution to Maxwell's second divergence equation (2.1b),[9] and where the constant of integration has been assumed zero[10] and where the imaginary part of any motion or field is a coupled orthogonal motion or field (see Appendix A). Here this particular coupled solution is consistent; the E- and H-fields can both be represented as spinors related by a similar radial dependence that solve the divergence equations and the first curl equation (2.1a–c). We continue with the time derivative of the H field. The cycle of differentiation implicit in Maxwell's equations is now completed to see if the form is consistent with the original trial form (2.2). In order to see if the resulting analytic form is the same as when we started our theoretical cycle, we take the curl of the time derivative of the H-field (2.1d) and substitute $c^2 = 1/\mu_0\varepsilon_0$ to obtain

$$\nabla \times \frac{\partial H}{\partial t} = \frac{-1}{4\pi} \frac{c^2 q}{\sin^2 \theta r^4} e^{j\phi} \hat{r} + \frac{c^2}{2\pi} \frac{q}{\sin \theta r^4} e^{j\phi} \hat{\theta} \tag{2.4}$$

Differentiating (2.2) twice with respect to time where the dependence $\phi = \omega t$ is again assumed reveals its similarity in form to the first term on the right-hand side of (2.4):

$$\frac{\partial^2 E}{\partial t^2} = \frac{-1}{4\pi\varepsilon_0} \frac{\omega^2 q}{r^2} e^{j\phi} \hat{r} \tag{2.5}$$

Equation (2.4) gives the two scalar components of current density in (2.1d) differentiated with respect to time. Starting with the E-field in (2.2) its second derivative is given by (2.5) and has a similar form to the first term on the right-hand side of (2.4). Note that by starting with negative exponential forms for the spinors a similar negative solution can be formed.

At this stage Maxwell's two divergence equations have supplied our initial spinorial forms while Maxwell's two curl equations have been cycled through and returned with a consistent E-field solution. For the hydrogen atom to be stable as detailed in Appendix B the H-fields need to rotate in the negative direction to the E-fields; then the electric and magnetic potential energies act against each other and can balance each other. As an aside note that if the charges on

---

[9] This H-field form can be derived from the vector and scalar potentials $\vec{H} = \frac{1}{\mu_0} \nabla \times \vec{A}$ and $\frac{\partial A}{\partial t} = \frac{-1}{\varepsilon_0}(E + \nabla\varphi)$.

[10] This constant cancels out in applications to periodic equilibria, but not for radiation reactions.

the two particles are repulsive, equilibrium can again be achieved if the particle spins allow the energies to be of opposite sign. This is a demonstration of the phenomenon of anti-particles known from particle physics. In the above analysis there was no spatial phase shift assumed. The only phase shift $\pi/2$ was due to differences between the E-field and the H-field and the differentials involved in Maxwell's equations. These phase shifts by themselves do not allow us to have positive and negative exponentials for our spinors. The presence of *both* signs allows internal standing waves as is known from the study of waveguides. We can achieve a mixed exponential form for the spinors if there are spatial phase shifts to provide an overall phase change of $2m\pi$, where m is the azimuthal mode. In the case where the orbital and cyclotron separations both provide phase shifts of $\pi/2$, we end up with a phase shift 2π across the analytic cycle. The overall self-field system starts with the E-field of the first particle; this provides the E-field to the second with a spatial phase shift of $\pi/2$ between the E-fields; the two H-fields are similarly spatially shifted in phase. The shift of 2π is over the analytic field cycle (length 4) $(\tilde{E}_1, \tilde{H}_1, \tilde{H}_2, \tilde{E}_2, \tilde{E}_3, \tilde{H}_3, \ldots)(0, -\pi/2, -\pi, -3\pi/2, -2\pi, -5\pi/2, \ldots)$. This reveals the resonance theory behind SFT in both time and space. Modes exist with any integer multiple of phase length. This resonance introduces two new quantum numbers into atomic physics that depend upon the binding energy; controlled by the constitutive parameters μ_0 and ε_0. In Chapter 3 this resonance is used to determine the photon mass by balancing the photon binding energy to that of an electron within a cycle of the atom. The analytic cycle also reveals a unique link between a photon stream and a particular particle pair. This coupling of a particular field stream to an electron and proton pair should allow analytic simplification of many-electron atoms.

The E- and H-fields form a cyclic stream of photons that move in two orthogonal planes. The fields can be separated into two coupled spatial coordinate systems (Fig. 2.5). In terms of these coordinates, a radial E-field spinor rotates in ϕ_o; this spinor and its time derivatives form a discrete cyclic set (of period 4 as with the phase shifts) in the $r_o - \phi_o$ plane (Fig. 2.5). An H-field spinor derivative also rotates in ϕ_o; the H-field spinor and its derivatives also form a discrete cyclic set of period 4 in the $r_o - \phi_o$ plane causing the electron to

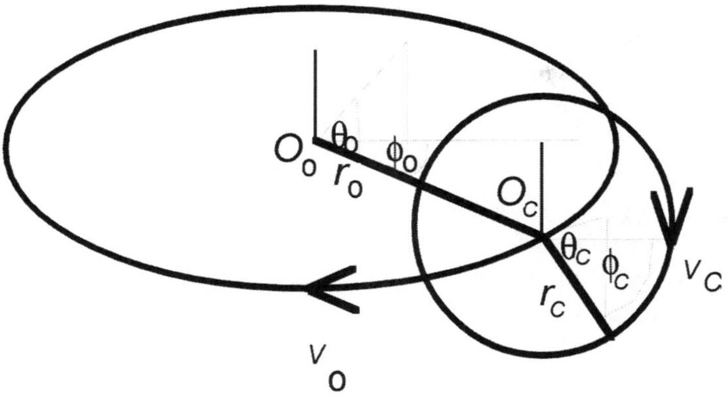

Figure 2.5. Coupled spherical coordinate system at the electron's electrical and magnetic centres of motion, coupled via the ML equations.

perform a cyclotron motion in the $r_c - \theta_c$ plane. The H-field cyclotron field spinor rotates in θ_c. Only four of the six spatial coordinates are independent, as in space–time coordinate systems.

Thus, given certain conditions the coupled spinor motions between two particles can be a self-field solution for the hydrogen atom (see Figures 2.5–6). The two spinors move at all times relative to their respective centres of motion. The motions depend on the ML equations and are orthogonal. In general for systems where

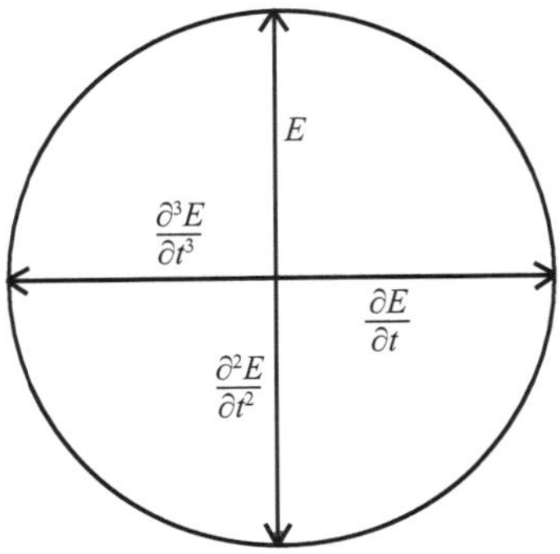

Figure 2.6. Time and space derivatives of the EM field form geometric series $n = 0, 1, 2, 3$, that return to the initial direction after a cycle of period 4. E-field series lies in the $r - \phi$ plane.

there may be many particles in an atomic structure, Maxwell's two curl equations provide three scalar equations per particle. For the Lorentz forces to be counterbalanced equation (2.1e) resolves into E- and H-field components providing two coupled scalar equations per pair of particles. So there are eight $(3 + 3 + 2)$ equations per pair of particles and the system of self-field equations is solvable there being four unknowns, the orbital and cyclotron radii and frequencies, and four equations per particle. This corresponds with the quantum numbers observed spectrographically in the hydrogen atom, the four-vectors of relativity and QFT in general.

Examining (2.4), a condition for the EM self-field solution to cycle theoretically is $v^2/\sin^2\theta = c^2$ where $v = \omega r$. Where $v = c$, a precondition is $\sin\theta = 1$ or $\theta = \pi/2$. Thus the rotations occur in the equatorial elevation. To this point the ML equations (2.1)–(2.5) describe the fields regardless of their application and involve the speed of light. Certain systems of charged particles are indeed self-sustaining and involve a rotation speed equal to the speed of light. We assume our equations are quite general describing atomic systems, or the internal structure of the photon, in fact anywhere self-fields occur. Our focus here is to see if these equations yield motions for the electron in the hydrogen atom. The electron's rotational speed is known to be around 1/137th that of light, $v_e = \alpha c$ where α is the fine structure constant. The condition for an equatorial elevation angle in the hydrogen atom is assumed as $\sin\theta = v_e/\alpha c$, so $\theta = \pi/2$ where $v_e = \alpha c$.

2.4 The Hydrogen Atom: Principal Mode

EMSFT proposes a bi-spinorial solution to Maxwell's equations capable of modelling atomic systems composed of charged particles. The theory is specialized to an electron moving dynamically inside the hydrogen atom in its principal mode. As detailed in Appendix A, the proton is assumed to have an infinite mass that complements the electron's motion electrically and magnetically (Fig. 2.2). Equatorial spinors $(\theta = \pi/2)$ represent the two rotational motions of the electron, an orbital spinor $\sigma_o(r_o, \omega_o) = r_o e^{j\omega_o t}$ associated with the E-field, and a cyclotron spinor $\sigma_c(r_c, \omega_c) = r_c e^{j\omega_c t}$ with the

H-field. Closely related to the motions of the electron, the fields too are spinors. The ML equations yield solutions for the electron representing a balance of its kinetic and potential energies.

It is now assumed that the spinor forms (A.15–16) yield the fields, forces, and motions required for dynamic equilibrium. These spinors provide COM E-and H-fields consistent with Maxwell's equations. The coupled fields balance the forces of the proton and electron. The orbital and cyclotron fields control the motions of the proton and electron and vice versa, the particle motions cause the fields that in turn cause the particle motions. A substitution the speed of light, $\varepsilon_0\mu_0 = 1/c^2 = 1/r_c^2\omega_c^2$ into (2.2) and (2.3) yields

$$E\left(\omega_o, r_o\right) = \frac{1}{4\pi\varepsilon_0}\frac{q}{r_o^2}e^{j\omega_o t}\hat{r} \tag{2.6}$$

and

$$H\left(\omega_c, r_c\right) = \frac{\omega_c}{4\pi}\frac{q}{r_c}e^{j\omega_c t}\hat{\phi} \tag{2.7}$$

where ω_o and ω_c are orbital and cyclotron angular velocities, and r_o and r_c are orbital and cyclotron radii. The spinorial field forms are consistent with the COM forces discussed in Appendix A. the motion of the electron forms an EM self-field solution, its position at any time is the sum of two spinors forming a bi-spinorial motion $\sigma_o(r_o, \omega_o)$ and $\sigma_c(r_c, \omega_c)$:

$$r\left(r_o, \omega_o, r_c, \omega_c\right) = r_o e^{j\omega_o t} + r_c e^{j\omega_c t} \tag{2.8}$$

In (2.8) each spinor refers to a COM; the orbital (E-field) COM is stationary, the cyclotron (H-field) rotates. This results from the coupled nature of the two spherical coordinate systems shown in Figure 2.5. Figure 2.7 shows an actual motion that compares with the probability densities of quantum theory. In terms of mathematical forms that lead to discrete or decoupled azimuthal modes, the complex exponential form of (8) is one such form. Solutions that return to their starting point, in other words are periodic, can maintain dynamic motions without net efflux or influx of energy. In general the various azimuthal modal forms that are written $e^{jm\phi}$ where $m = 1, 2, 3\ldots$ are a prerequisite to a discrete or quantum physics.

For the simplified case of the principal mode, a system of equations based on Maxwell's curl equations (2.1c–d) can be

Figure 2.7. Plot of bi-spinor solution in $r_o - \phi_o$ and $r_c - \theta_c$ planes, where $r_o = 4r_c$ and $\omega_c = 16\omega_o$.

rewritten where the orbital and cyclotron frequencies are assumed equal $|\omega| = |\omega_o| = |\omega_c|$. The $e^{j\omega t}$ factor also simplifies out, the system reduces to three equations in three unknowns (ω, r_o, r_c):

$$\frac{q}{4\pi\varepsilon_0 r_o^3} + \frac{\mu_0\omega^2 q}{4\pi r_c} = 0 \tag{2.9}$$

$$\frac{c^2}{2\pi}\frac{q}{\omega r_c^4} = \frac{q}{2\pi r_c^2}\omega \tag{2.10}$$

$$\frac{1}{4\pi}\frac{c^2 q}{\omega r_o^4} + \frac{1}{4\pi}\frac{\omega q}{r_o^2} = \frac{-q}{2\pi r_o^2}\omega \tag{2.11}$$

Equations (2.9)–(2.11) follow substitution of the field spinors (2.6) and (2.7) into (2.1c) and (2.1d), respectively. Equation (2.9) follows (2.1c), while (2.10) and (2.11) are the two components of current performed by the electron, a negative sign in the source terms of (2.11) indicates anticlockwise current. Equations (2.9)–(2.11) are satisfied if as well as the radii and angular frequencies having equal magnitude, a phase difference of $\pi/2$ exists between ω_o and ω_c or equivalently $\pm j$. The potentials oppose each other in (2.9).

By substitution of $\varepsilon_0\mu_0 = 1/r_o^2\omega_o^2$ and cyclotron area constants $s_o = \pi r_o^2$ and $s_c = \pi r_c^2$, (2.9)–(2.11) yield three equations in three unknowns (ω, r_o, r_c):

$$\frac{1}{4\pi}\frac{\omega^2 q}{\varepsilon_0 r_o} - \frac{1}{4\pi}\frac{\omega^2 q}{\varepsilon_0 r_c} = 0 \tag{2.12}$$

$$\frac{c^2}{2\pi}\frac{q}{r_c^4} = \frac{q}{2\pi r_c^2}\omega^2 \tag{2.13}$$

$$\frac{c^2}{4\pi}\frac{q}{r_o^4} + \frac{c^2}{4\pi}\frac{q}{r_o^4} = -\frac{1}{2s_o}q\omega^2 r_o \tag{2.14}$$

What are termed the field terms are on the left-hand side and the particle or source terms are on the right-hand side of (2.12)–(2.14) as in (2.1c–d). As will be seen, quanta of Planck's number, photons, are found theoretically on the right-hand side and the electron parameters on the left-hand side, vice versa to the normal particle-field system, indicating a two-way balance between coupled field and particles.

2.5 Forces and Energies of the Principal Mode

Equations (2.12)–(2.14) can be written in terms of the forces and energies associated with them. Studies of electrostatic and gravitational orbits reveal that centripetal and centrifugal forces can establish balance while performing circular or elliptic motions. Where an electron and a proton are in dynamic equilibrium, pairs of electrical and magnetic forces each balance the mass of the particle in orthogonal directions of motion. Equation (2.15) can be written for the electron's orbital motion. Substitution of the E-field spinor (2.6) and the orbital position spinor $\sigma_o(r_o, \omega_o)$, (2.8) yields an electric form of the virial theorem

$$m_e\frac{d^2 r_o}{dt^2} = qE \tag{2.15}$$

$$m_e\frac{d^2 r_o}{dt^2} = \frac{1}{4\pi\varepsilon_0}\frac{q^2}{r_o^2}e^{j\phi} \tag{2.16}$$

$$\frac{1}{2}m_e\omega_o^2 r_o e^{j\omega_o t} = \frac{1}{8\pi\varepsilon_0}\frac{q^2}{r_o^2}e^{j\omega_o t} \tag{2.17}$$

$$\frac{1}{2}m_e\omega_o^2 r_o^2 = \frac{1}{8\pi\varepsilon_0}\frac{q^2}{r_o} \tag{2.18}$$

where m_e is the mass of the electron. The charges being equal and opposite, the E-field is attractive and the magnitude of the orbital centripetal force is $m_e\omega^2 r_o$. The virial theorem given by (2.18) links the kinetic and potential energies due to the E-field. How can the electron and proton maintain a constant E-field spinor? The two rotate around a common stationary centre of motion constant orbital separation is achieved if the two maintain a certain prescribed angular velocity in ϕ and if the phases of their ϕ rotations differ by 0 or π, where the larger separation is the lower energy state. On its own such an orbital system would be unstable.

Equation (2.19) can be written for the electron's cyclotron motion. Substitution of the H-field spinor (2.7) and the cyclotron position spinor $\sigma_c(r_c, \omega_c)$, (2.8) yields a magnetic form of the virial theorem

$$m_e \frac{d^2 r_c}{dt^2} = q v_c \times B \tag{2.19}$$

$$m_e \frac{d^2 r_c}{dt^2} = \frac{-\omega^2 \mu_0 q^2}{4\pi} e^{j\omega_c t} = \frac{1}{4\pi\varepsilon_0} \frac{q^2}{r_c^2} e^{j\omega_c t} \tag{2.20}$$

$$\frac{1}{2} m_e \omega_c^2 r_c e^{j\omega_c t} = \frac{-1}{8\pi\varepsilon_0} \frac{q^2}{r_c} e^{j\omega_c t} \tag{2.21}$$

$$\frac{1}{2} m_e \omega_c^2 r_c^2 = \frac{-1}{8\pi\varepsilon_0} \frac{q^2}{r_c} \tag{2.22}$$

where $\hat{r}_c = j\hat{r}_o$ and the charges due to the electron and proton rotate in the same direction and hence the cyclotron force $m_e\omega^2 r_c$ is repulsive (centrifugal). Again, a constant cyclotron radial separation can be maintained provided both particles maintain a certain angular velocity in θ, and their θ rotations differ in phase by π (for the lower state). Such a cyclotron system would be unphysical as well as unstable on its own requiring the E-field for its existence which in turn requires the H-field; together the fields form an "entrapped" EM self-field inside the atom.

Using the virial relationships (2.18) and (2.22), (2.12)–(2.14) can now be rewritten.

$$\frac{1}{4\pi\varepsilon_0} \frac{\omega^2 q}{r_o} - \frac{1}{4\pi\varepsilon_0} \frac{\omega^2 q}{r_c} = 0 \tag{2.23}$$

$$m_e v_c^2 = 2\hbar\omega \tag{2.24}$$

$$m_e v_o^2 = 2\hbar\omega \tag{2.25}$$

Note that for (2.24) and (2.25), $2\hbar\omega = \frac{q^2}{8\pi^2\varepsilon_0}$ is known at this stage before the solution is determined analytically. Hence discrete quanta of Planck's energy are found on the right-hand sides of (2.24) and (25). These are source terms that can be traced back from the right-hand side of (2.1d). Further a connection to quantum theory is obtained from (2.18) and (2.22) by dividing (2.18) by ω_o and (2.22) by ω_c, so $\hbar = \frac{q^2}{8\pi\varepsilon_0 v_0} = \frac{q^2}{8\pi\varepsilon_0 v_c}$. In this form, Planck's constant is a variable of motion, dependent on the solution of the equations. The quantum nature of the electron's motion has long been observed. The system of three equations in the unknowns ω and r_o and r_o are based on classical energy forms. Eliminating (2.24), the system may be recast in terms of the energies in diagonal matrix form

$$\begin{bmatrix} 1 & 0 \\ 0 & 1 \end{bmatrix} \begin{bmatrix} V \\ T \end{bmatrix} = \begin{bmatrix} 2\hbar\omega \\ \hbar\omega \end{bmatrix} \tag{2.26}$$

In (2.26), $V = V_0 = \frac{1}{4\pi\varepsilon_0}\frac{q}{r_o} = V_c = \frac{1}{4\pi\varepsilon_0}\frac{q}{r_c}$, and $T = T_0 = 1/2m_e v_o^2 = T_c = 1/2m_e v_c^2$ are the orbital or cyclotron components of the potential and kinetic energies. In this form $\omega = |\omega_o| = |\omega_c|$ and $r = |r_o| = |r_c|$. From inspection (2.26) agrees with the virial theorem. Notice that the left-hand side of (2.26) contains the elements needed to rewrite the Schrödinger equation for the atom in SFT form.

2.6 Analytic Solution for Infinite Mass Proton

The complete matrix equation forms two conjugate parts, for the orbital and cyclotron motions, each a sub-matrix containing two equations. The principal mode case has simplified the analysis. Equation (2.26) can be solved analytically and can be compared with the expressions in Bohr's theory for the Bohr radius, resonant frequency, and Rydberg's number

$$r_o = r_c = \frac{4\varepsilon_0\hbar^2}{\pi m_e q^2} = 4r_B \tag{2.27}$$

$$\omega_o = \omega_c = \frac{\pi q^4 m_e}{16\varepsilon_0^2\hbar^3} \tag{2.28}$$

$$R_o = R_c = \frac{q^4 m_e}{32\varepsilon_0^2 \hbar^3 c} = R/4 \tag{2.29}$$

From (2.27) the orbital and cyclotron radii are $4r_B$ where, as in Appendix B, $r_B = 0.5291771 \times 10^{-10}$ m. From (2.28), the orbital and cyclotron angular frequencies are $\omega = 1.033532 \times 10^{16}$ rad/s. If the two components of energy from the electron's motion are combined with two equal components of energy from the proton's motion, we obtain the frequency $f_e = 6.579671 \times 10^{15}$ s^{-1}, or a wave-number 10,973,710. In this way the total system energy is linked to the motions of both the electron and proton; as the electron changes state, the motion of the proton complements any such change.

2.7 Complete Azimuthal Mode Solution: Balmer Formula

At this stage the study has revealed that there are two spinors associated with the electron's motion in the principal mode that have equal magnitude and are separated in phase by $\pi/2$. The complete eigensolutions will involve two radial modes and two frequency modes. The radial modes are required to be periodic; an equation based on (25) where $v_m = \omega mr$ and $v_n = j\omega mr$ can be written.

$$1/2 m_e v_m^2 - 1/2 m_e v_n^2 = 2\hbar\omega \tag{2.30}$$

Algebraic manipulation of (2.30) gives the analytic Balmer formula (as distinct from the Balmer series), which is based on the energy emitted if an atom shifts between two radial modes.

$$\nu_{mn} = R\left(\frac{1}{m^2} - \frac{1}{n^2}\right) \quad \text{where } m^2 = 1, 2, 3, \ldots \text{ and } n = 2, 3, 4, \ldots \tag{2.31}$$

There are other possible radial modes whereby a "fine-structure" is revealed. A complete series analysis reveals angular modes that are much closer to each other than the radial modes. If the principal mode analysis is repeated using higher order separated azimuthal modal series for both E-field ($e^{jm\phi}$) and H-field ($e^{jm\phi}$) spinors a complete azimuthal series is easily obtained from (2.4) and (2.5):

$$\nabla \times \frac{\partial H}{\partial t} = \frac{-1}{4\pi} \frac{m^2\omega^2 q}{r^2} e^{jl\phi}\hat{r} + \frac{m^2\omega^2}{2\pi} \frac{jq}{r^2} e^{jl\phi}\hat{\phi} \tag{2.32}$$

$$\frac{\partial^2 E}{\partial t^2} = \frac{-1}{4\pi\varepsilon_0}\frac{n^2\omega^2 q}{r^2}e^{jk\phi}\hat{r} \tag{2.33}$$

where it is assumed $\sin\theta = 1$ at equatorial elevations and once again the speed of light is not proscribed. The energy levels associated with the azimuthal modes are separated by a quantum factor $\alpha\hbar\nu$ involving the fine structure constant α. Overall there are four modes arising from the two spinors. These modes correspond to the experimental data provided by spectroscopy and theory provided by quantum theory.

2.8 Stability

An important question concerns the stability of the EM spinor solution given the well-known instability of classical EM theory (see Appendix A). The total energy exhibits non-trivial stability points in phase space and may be written to emphasize the nature of the energy forms present:

$$\langle E_{Tot}\rangle = 1/2m_e\left|v_o^2\right| + 1/2m_e\left|v_c^2\right| + \frac{1}{4\pi\varepsilon_0}\frac{\omega_o^2 q}{r_o} - \frac{1}{4\pi\varepsilon_0}\frac{\omega_c^2 q}{r_c} \tag{2.34}$$

The total energy for the electron includes the photon field energy and is balanced when $V_o = V_c$, and $T_o = T_c$. The sign associated with the magnetic energy shows that EMSFT is consistent with the negative magnetic moment associated empirically and theoretically via QED with the electron (Landé g-factor). In terms of ML equations, (2.1c) is a balance of electric and magnetic potential energies and (2.1d) a balance between electric and magnetic kinetic energies and the total energy of the photonic binding field.

In Appendix A the long-asked question of why the electron in the atom does not spiral inwards as it emits energy is answered in terms of the coupled self-fields that are effectively swapped by the two particles, the electron and the proton. In hindsight this appears in some ways obvious yet this same issue remained unanswered during the greater part of the 20th century. Had the self-fields been understood as a mutual effect our scientific history and all that this entails may have been vastly different during the 20th century.

2.9 The Proton's Motion in the Hydrogen Atom

If the proton has "infinite mass" (Figure 2.2), it is assumed not to move from the centre-of-mass and the electron moves with the entire orbital and cyclotron radius. If instead a finite-mass proton is used (Figure 2.8), the EMSFT equations involve both the electron and the proton. In this case, the electric and magnetic potentials become functions of the orbital and cyclotron radial distances of both electron and proton. There are six equations where, as in (2.23)–(2.25), the cyclotron and orbital frequencies are assumed equal. From these equations, Rydberg's number can be adjusted giving $R_H = \frac{q^4 m_e}{8\varepsilon_0^2 \hbar^3 c} \frac{m_p m_e}{m_p + m_e} = 10967757.78\ \mathrm{m}^{-1}$, which is in excellent agreement with spectroscopic data $10967757.6\ \mathrm{m}^{-1}$. The proton orbital and cyclotron radii can be estimated as $4r_B \frac{m_e}{m_p+m_e} = 1.169634 \times 10^{-14}$ m while the electron's orbital and cyclotron radii become $4r_B \frac{m_p}{m_p+m_e} = 2.115556 \times 10^{-10}$ m. These results are similar to the concept of *reduced mass* $\mu = \frac{m_p m_e}{m_p+m_e}$ well known in Bohr's early work. This estimate of the size of the proton is much bigger than current estimates from scattering experiments of ~ 1.2 fm and is probably due to the fact that there are no strong nuclear forces in our model. The proton's motion can be better estimated if the strong nuclear forces are explicitly included using a more refined three-quark model (Figs. 2.9–2.10). In the model shown in Fig. 2.10 both EM photons and strong nuclear gluons are present.

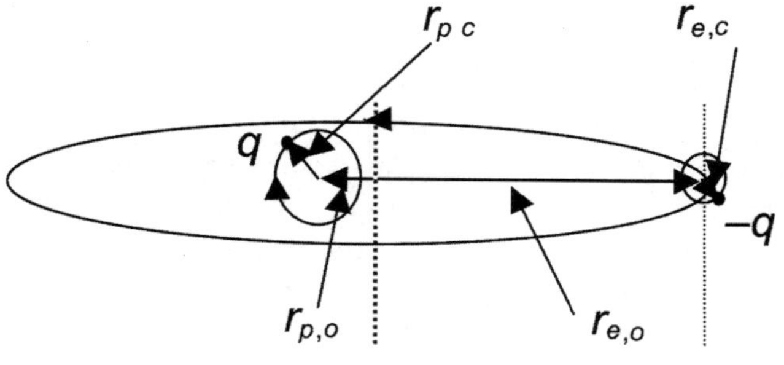

Figure 2.8. Hydrogen atom modelled with finite-mass proton.

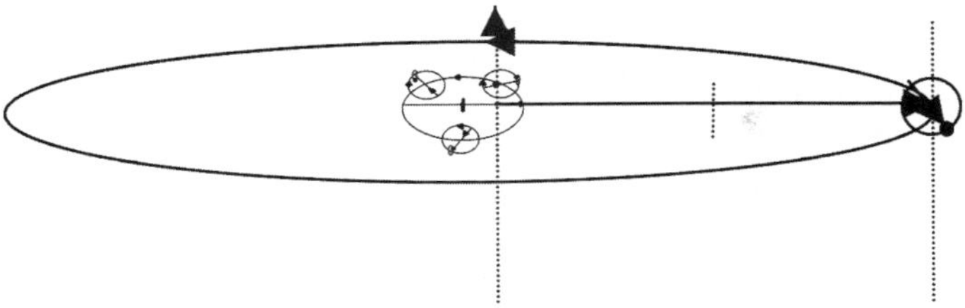

Figure 2.9. Hydrogen atom modelled with both strong and EM fields with a finite-mass proton consisting of three quarks coupled via gluon fields.

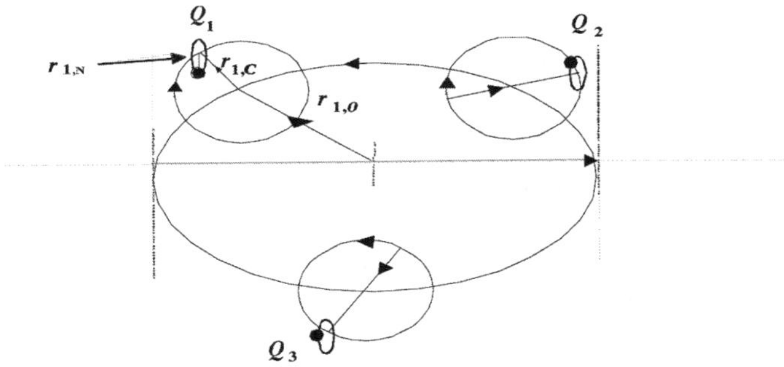

Figure 2.10. Proton modelled as three quarks coupled via gluon fields.

2.10 Comparison of EMSFT Solution with Other Field Solutions

Compared to classical and quantum fields, EMSFT fields inside the hydrogen atom are very different and physically realistic. They do not exist in all space, only inside the atom; the radiating fields inside the atom have been balanced. The EMSFT fields are assumed to consist of discrete photons. The fields are like "beads on a spiral-helical string" in some ways reminiscent of the helical shape of the macromolecule DNA. Even inside the atom, photons do not occupy the whole interior of the atom; photon streams exist only between the proton and electron whose motions are also bi-spinorial. In EMSFT there is no difference in classification between field and particle, both yield spinorial solutions to ML equations.

The classical field vectors are measured macroscopically with dipoles and coils and therefore are the fields of the everyday world we live in, but these are *not* the COM fields that are used in this report. These EMSFT fields are not measurable although they *do exist* in the real world and can be modelled as spinors. There are many possible pairs of Lorentz compatible E- and H-fields,[11] but only the COM fields are invariant in form as velocity is varied. As the rotation speed varies no relativistic correction terms are needed. The COM E-field is *not* a point-to-point field as in Coulomb's law, but is taken between the centres of motion of *both* the electron and proton, two points in vacant space. These fields cannot be measured directly in the real world but can be calculated.

Comparison of SFT and quantum theory shows there is an intimate relationship between these theoretical field models, both capable of modelling the hydrogen atom. The spinor equations, (2.12)–(2.14), and energy equations (2.23)–(2.25), in particular the relationship between their right-hand sides and Planck's energy quanta demonstrates the actual basis of QED and QCD. QFT relies on a mixed formulation of two conjugate potentials enabling observables such as energy shifts between allowed states to be determined. Compared to this, the spinors of EMSFT provide distance and frequency as separated entities. Therefore, unlike QM that cannot provide deterministic motions but must provide probabilistic solutions, there is no uncertainty in EMSFT. In EMSFT motions and frequencies are defined allowing comparison with data such as the probability densities provided by QM by choosing a suitable time sample interval (see Fig. 2.7). In stark contrast to the edicts of HUP the actual time-evolution of the motions can be obtained from EMSFT.

2.11 Planck's Constant

Empirically the value of $\hbar$ comes from (2.24)–(2.25), where $2\hbar\omega = \frac{q^2}{8\pi^2\varepsilon_0}$, so $\hbar = \frac{q^2}{8\pi\varepsilon_0 v_0} = \frac{q^2}{8\pi\varepsilon_0 v_c}$ can be calculated from the solution of the Bohr radius and the resonant frequency (2.27)–(2.28). To see

[11]All such Lorentz-compatible vector pairs form a set of triangles lying on the circumference of a semicircle whose diameter squared is proportional to the energy density.

the relationship between Planck's constant and velocity let $\hbar = \frac{E_0}{\nu_0} = \frac{q^2}{8\pi\varepsilon_0 r_0}\frac{1}{\nu_0} = \frac{q^2}{8\pi\varepsilon_0 v_o}$ where ν_0 is the frequency and E_0 is seen as E-field potential energy. Since $v_0 = r_0\omega_0 = v_c = r_c\omega_c = 5.469222 \times 10^5$ m/s, we can compare with the known value of Planck's constant (Appendix B) with our estimate to an accuracy of 7 significant figures.

Obviously this raises an issue at the heart of quantum theory. Condon and Odabasi discuss the acceptance of Planck's constant in the pre-quantum era, 1900, as a fundamental concept in that quantities associated with the description of atomic phenomena assumed only a discrete set of allowed values that incorporate Planck's constant. After failing to find a classical framework for the thermodynamic equilibrium of matter, otherwise known as *blackbody radiation,* Planck in an "act of desperation" solved the problem by assuming only certain analytic frequencies could exist for the radiation. At that time Planck and his contemporaries thought of the solution as a mathematical artifice and not a defining point in the evolution of quantum physics. He numerically solved a two-variable problem in h Planck's constant, and k Boltzmann's constant, using empirical data and thus established the groundwork for the photoelectric relationship $E = \hbar\nu$ that Einstein verified in 1905 via the photoelectric effect.

SFT does appear to be the theoretical framework that Planck and many others including Einstein sought. That Planck's constant represents a variable in EMSFT demonstrates the fundamental nature of SFT. This finding follows from the assumption that the speed of light is not proscribed from being variable in EMSFT. Empirical measurements, no matter how widespread their physical reach, cannot determine whether any quantity, no matter how cherished, is a constant or variable until a theoretical framework has been established. EMSFT is an eigenvalue problem and its solution consists of four eigenvalues. The constitutive parameters of free space, ε_0, the permittivity, and μ_0, the permeability, are also called constants where deep space is being considered. These parameters are constants where the universe is considered homogeneous and isotropic. In reality, there are variations, for example the outer boundary of the universe, or inside the nuclei of atoms. This means there are six degrees of freedom in total in EMSFT, where the fields are EM, leaving aside nuclear forces and weak gravitational forces.

Where does the self-energy of the system come from? The theory suggests that photons supply the energy to the electron and the proton and vice versa in the form of elastic collisions. That discrete quanta of Planck's energy are involved in the particle–field interactions strongly implicate the photon as the energy provider/mediator for the atomic system. The dynamics of the atomic system can be written as (2.19)–(2.21), an inhomogeneous system where the right-hand side contains the photon energy. In addition to representing Planck's quanta, these are derived directly from the current densities of Maxwell's equations (2.1c). This suggests two interfacing systems of motion at differing scales each supplying the other's energy; two interfacing inhomogeneous systems of equations. The stability of the field/particle system, (2.32), may also be reflected in a similar equation as seen from the perspective of the photon, hence a state of balance between the motions and interactions of the proton, electron, and photons may exist.

That Planck's constant exists on the right-hand side of (2.24) and (2.25) demonstrates empirically that the source terms of the E- and H-fields of the electron are indeed photons as is the case for QFT. Looking at the matrix structure of EMSFT, the fields (photons) represent the source terms on the right-hand side for the electron's motion. Vice versa, the motion of the electron (and proton) can be written as source terms for the photons if they were written as the particles. The "fields" and "particles" are thus interchangeable in the mathematics of EMSFT. The source term variable $\hbar = \frac{q^2}{8\pi\varepsilon_0 v_0} = \frac{q^2}{8\pi\varepsilon_0 v_c}$ on the right-hand sides in this case will be identical in both cases yielding Planck's energy. A most fundamental finding indeed and an insight into self-fields and their balance inside atoms. Not only are the particles, in this case the proton and electron, locked in a dynamic balance but the photons are demonstrably part of this balance. As well as the electron, an EMSFT application concerning the internal dynamics of the photon can be examined.

The E- and H-fields of classical physics are continuously differentiable "objects" in space–time, as waves moving through ubiquitous *ether*. In EMSFT, the fields are represented as photon *streams* that cycle with the particles that rotate. Thus EMSFT is not a classical field theory, but the ML equations and a *quantization* are incorporated. In EMSFT quantization is not heuristic but is part of the solution. There are two photon streams, from proton to electron and electron to proton, flowing at a constant rate (photons per second) forming one cycle. Each photon's motion is similarly bi-spinorial to that of the electron and proton but of different magnitude and overall shape. Whereas the electron or proton circles along its orbital path, the photons move out from the proton to the electron or vice versa in a cycle of the binding energy. The photons and particles move in their near fields as an intricate but balanced standing wave.

2.12 SFT and Its Applications to Other Physical Phenomena

In contrast to the mathematical use of unit spinors within classical field theory and QFT, SFT uses the spinor as a function to model physical motion. As such the spinor can be applied as a variable across physics. Particle motions inside high-energy accelerators, orbital and cyclotron spin motions in solar systems and galaxies, and the double-helix structure of DNA exemplify the existence of spinors in the real world. We may think here of numerical modelling and the use of approximation functions including splines that can be used for modelling statistical data. SFT can be applied to the fields and particles within the photon, the nucleus, to the motions of solar systems and super galaxies. There appears to be a hierarchical system of physical spinors that can be modelled by SFT. Spinor fields undergo a form of "differentiation" on scales larger than the atom and a form of "integration" on smaller scales (discussion on quarks below), while the spinor particles appear to operate in the opposite direction of scale. Note that we do not use the term "mass-particles" because it seems from SFT that both fields and particles both have mass. The picture emerges from SFT of coherent elastic collisions that bind fields and particles together.

For dynamic balances where elliptic solutions are valid the spinors vary over each rotation and a numerical approach may be needed to obtain the motions. Further, atoms are not stable in only the one state as given by EMSFT in the solution previously discussed since the equilibrium points are saddle points allowing for small perturbations around a central equilibrium state without becoming unstable. In the sense that there are many possible motions involved in the overall spinorial motion, the motions may average out to the simple circular solutions as solved. In EMSFT, the fields, forces and motions, can all be modelled as spinors. EMSFT also applies to non-equilibrium applications where the spinors can vary with time. As we will see in Chapter 3, the spinor motion can also be modelled as a polygon involving elastic collisions between the field and particle. As noted in Chapter 1, the DNA molecule suggests a piecewise linear spinor.

The helical–spiral motion of the photon in the hydrogen atom is not unlike spiral forms seen in solar systems and galaxies, and there appear to be larger gravitational structures such as super clusters within our own Universe. There is a growing number of physicists who consider that there may exist multiverses. The spiral shaped photonic solutions of the hydrogen atom have half-integer "spins" associated with them, 180°, 540°, 810° etc. reminiscent of the angular spin states of quantum theory whereby dynamic balances in atoms can be achieved. One particular series of dynamic balances relates to the successive radial gradients of the EMSFT. As outlined in Table A.1 and investigated in Section 2.12 dipolar EM fields, both electrical and magnetic, provides a possibly four tiered model for the gravitational forces observed within the Universe (see Chapter 3).

Not unlike string theory, photon streams are a feature of the binding energy within the hydrogen atom. In the hydrogen atom there are helical–spiral streams of photons, one flowing from proton to electron, and another from electron to proton. These two streams combine to form a cycle that maintains motion and stability. The two streams associated with each particle are orthogonal to each other and hence so too are the E- and H-forces, while the photons from each particle are anti-parallel to those of the other particle. The concept of a photon beam or stream is an accepted physical description. Instead of a straight-line stream where all particles have the same velocity, the stream is a helix where the photon velocities have a circular variation. These streams support and create the spinor motions of the electron. The two mutual photon streams also form coupled spinor motions. The spinors formed by the motions of charged particles and the photons are valid solutions to the ML equations; with the motions of the fields and the particles coupled together in an intricate balance. Barut discussed the duality of particle and field in classical electromagnetics and QFT.

The EM field studied by Einstein in the early 20th century consisted of discrete photons. In classical field theory a flux of photons is assumed to radiate equally in all directions, over all solid angles to infinite radius. In the EMFST, photon streams do not radiate in all directions but only between masses; further they do *not* radiate to infinity since they are not massless. These distinctions are crucial in solving current cosmological problems seen to require a quantized form of gravitation. One of the insights of SFT is that a discrete physics depends on a SFT balance between unequal masses, while a continuous physics depends on a SFT balance between equal masses.

The coupled spinors come from the motions of the electron *and* the proton, a mutual motion. Quantum theory involves mathematical entities such as matrix or tensor elements. In EMSFT, the spinors are time-varying physical entities, motions, and can be used as the variables in the mathematical formulation. The spinors form elements of a closed, self-sustaining system composed of two particles and their binding field whose motions are all mutual and complementary. Each spinor can be written mathematically as column vectors that are coupled via 2×2 interaction sub-matrices of the overall interaction matrix. One of the coupled spinors is associated with the E-field as seen by the electron, and the other with its H-field. Conversely, the proton sees these two spinors as an H-field and an E-field, respectively. This symmetry is not observed in classical electromagnetics or QFT.

2.13 SFT Gravitational Model Based on Dielectromagnetics

The history of the mathematical study of gravitation began around the 1660s with Newton's discovery of the inverse-square law given by (2.36a), where Newton's gravitational constant $G = 6.67 \times 10^{-11}$ $\mathrm{Nm^2/kg^2}$. Newton's law presumed an instantaneous *action at a distance*. Shortly after this time a growing realization emerged that the speed of light was finite. In 1675 Rømer estimated its speed by observing the times taken for eclipses of the moons of Jupiter at different times of the year when Jupiter's distance from Earth could be calculated. In 1887 the Michelson–Morley experiment revealed the experimental invariance of the speed of light and in 1915 Einstein's presented his theory and validation of general relativity (GR). GR is considered to warp space–time and transform Euclidean frames of reference into curved space–time in which there are no forces, only geometric entities such as curvature and directional gradients. As Einstein pointed out, an observer sitting on the line of curvature experiences no forces. A photon "falls under gravity" along a geodesic curve. Found intuitively difficult to accept by many when announced, the experimental evidence of the precession of the perihelion of Mercury supported GR over the Newtonian model which could be expressed as a first approximation of GR. Light bends near massive objects and black holes form as interiors of stars of sufficient mass collapse, exhibited in GR as singularities within the geometry of space–time. Einstein's original field equation is written as

$$G_{\mu\nu} + \Lambda g_{\mu\nu} = \frac{8\pi G}{c^4} T_{\mu\nu} \tag{2.35}$$

where $G_{\mu\nu}$ is the Einstein tensor ($G_{\mu\nu} = R_{\mu\nu} - \frac{1}{2}Rg_{\mu\nu}$, where $R_{\mu\nu}$ is the Ricci curvature tensor), R the scalar curvature, $g_{\mu\nu}$ the metric tensor, Λ is the cosmological constant, G is the gravitational constant, c the speed of light, and $T_{\mu\nu}$ the stress energy tensor.

To give validation of GR Einstein calculated the advance in the perihelion for a planet in circular orbit and added a second-order correction for its elliptic orbit. The advance in distance for a circular orbit is $24\pi^3 r^3/c^2t^2$. The correction for elliptical motion is $r(1-\varepsilon^2)$. The advance in angle now becomes $24\pi^3 r^3/c^2t^2(1-\varepsilon^2)$ radians. This was solved as a geodesic variation problem.

A gravitational model can also be based on the electric and magnetic dipolar properties of matter that form a SFT binary (see Table A.1). We can see a similarity between GR and SFT in that at balance, the masses move along a geodesic that minimizes the forces upon them. The forces are balanced and the particle continues to *free fall.* As discussed in Chapter 3, this SFT model of gravity yields a structured model based on the different types of matter and bosons of particle physics that are seen to qualitatively match the gravitational structures seen in the Universe. The role of electromagnetics in this search for theoretical understanding of gravity is interesting because of the similarity between Newton's and Coulomb's inverse-square laws

$$F_{grav} = \frac{G\, m_e m_s\, \hat{R}_{SE}}{\left| R_{SE}^2 \right|} \tag{2.36a}$$

$$F_e = \frac{q_1 q_2}{4\pi \varepsilon_0} \frac{\hat{r}}{\left| r^2 \right|} \tag{2.36b}$$

While Newton and Einstein came up with mathematical equations to describe gravitation, neither model can explain why gravity works and what keeps gravity continuing to cause an attractive force. The rotation of Earth is put down to the conservation of angular momentum, which explains the continuance of Earth's spin about its own axis but not why it occurs.

The possible application of SFT to gravitational structure within the Universe is based on two factors: (1) as in EPR there is a photonic connection between all matter in the Universe discussed in some detail in Chapter 3, a remnant of the proximity of matter at the instant of the Big Bang, and (2) a dipolar EM model of matter. As shown in Table A.1 both diamagnetic and dielectric materials can radiate as they accelerate or change direction. A differential form of EMSFT is proposed to apply to gravitation at the solar system level, while other forms of matter and their bosons appear to apply to other gravitational structures seen within the Universe. In Chapter 3, we present a model for gravitational structures throughout the Universe based on the idea that mathematical differentiation of the EM field is at the heart of theoretical gravitation. Similar to EMSFT the action of a coupled field as a binding agent between dipolar EM matter is involved. There are various aspects of this phenomenon that are encountered in this proposed form of gravity and we examine a few in the following.

Shakespeare's Juliet asks "What's in a name? That which we call a rose by any other name would smell as sweet." Materials

Figure 2.11. Earth's rotation is the sum of two dielectromagnetic effects. (Credit: Wikemedia.) See also Color Insert.

known as *dielectromagnetics* are produced using new generation technology, which are composites of both dielectrics and magnetic materials. A range of such components promise benefits including production costs, energy efficiency, and product recycling. Epoxy resin and iron powder composites can be produced to give a range of sizes from nanoscale for biological use to macroscopic devices for use in the power industry. Some including the author have suggested atoms are dielectromagnetics. There is another scale of composite dielectromagnetic object, namely the Earth (Figure 2.11), in fact all the planets including the exoplanets, the extrasolar planets. While the term dielectromagnetics is commonly used to describe dielectric and magnetic composites, to SFT it is a form of EM theory describing the coupled photonic interaction between a sun and its planets. The electromagnetic waves Maxwell discovered to be coupled were found to radiate over oceans and space. Dielectromagnetic (DEM) waves are another form of coupled EM wave that can mediate gravitational forces over the huge distances of deep space. Similar to the way EMSFT solves for the eigensolutions of an atom, DEMSFT may solve for the macroscopic gravitational effects of atoms due to other atoms. The binding energy, a differential photonic energy density, may have a signature spectrum able to indicate Earth-like planets.

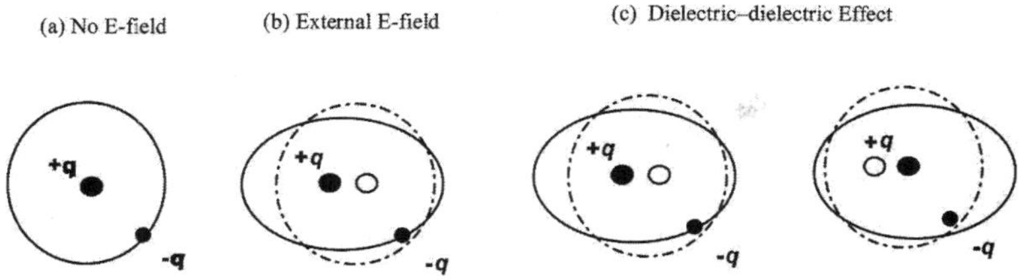

Figure 2.12. Polarization of the hydrogen atom results in an asymmetric shift of charge over the atomic volume due to the application of an external E-field. A non-resonant coupled dielectric effect occurs when two atoms affect one another. (If we rotate the ϕ-axis to match the major axis of the ellipse we again have an axisymmetry.)

The effects of B-fields upon atoms were studied in 1896 by Zeeman who examined the splitting of the spectral lines of atoms exposed to static B-fields. E-field effects on atoms were studied by Stark in 1913 who examined the changes to spectral lines of atoms exposed to static E-fields. Uhlenbeck and Goudsmit in 1927 discovered atomic spin, another form of splitting possessed by the electron not involving the nucleus. These are uncoupled effects on a single atom where the atom can be understood as a form of DEM particle. Shown in Figs. 2.12a,b and 2.13a,b the orbital and cyclotron rotations of the hydrogen atom are distorted by applied external E- and H-fields. These effects can be qualitatively understood via classical electromagnetics where the hydrogen atom can be seen to form both an E- and an H-field dipole, although QM is required to understand them quantitatively.

Shown in Figs. 2.12c and 2.13c similar dielectric and diamagnetic effects result when two atoms are brought into proximity of each other. These are non-resonant dipole–dipole effects. The two diamagnetic atomic dipoles rotate around each other similar to the flattening of a gyroscope. The basis of classical electromagnetics are the time-invariant laws of Coulomb and Ampere. Similarly there are two laws that apply to dipole–dipole interactions; these too are time-invariant and based on classical electromagnetics. There is a dipolar analogue of SFT that applies to the electron and proton of the hydrogen atom that applies to the atomic domain such as the two atoms shown Figs. 2.12c and 2.13c and to macroscopic

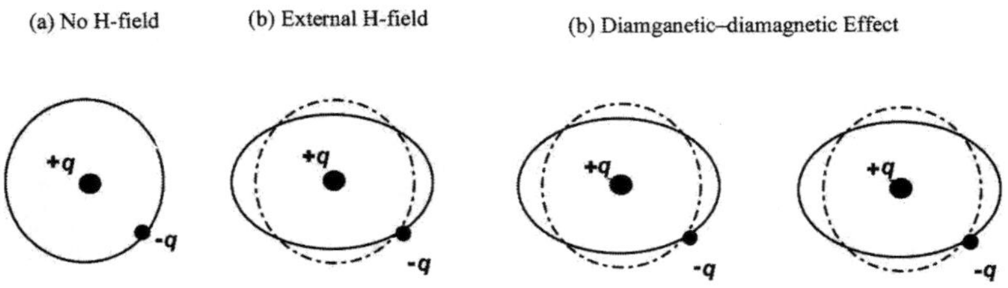

Figure 2.13. Diamagnetic flattening of the hydrogen atom results in an axisymmetric shift of charge over its atomic volumes due to the application of an external H-field. A non-resonant coupled diamagnetic effect occurs when two atoms affect one another.

conglomerates such as the Earth and the Sun. It is important to note that the effects shown in Figs. 2.12c and 2.13c are non-resonant. At resonance the mutual effects will be minimized and the two atoms form a stable hydrogen molecule. In the macroscopic case resonance is averaged over the atomic conglomerate while individual atom–atom forces are not. With a conglomerate there are many photon streams forming a macroscopic beam.

By taking the derivative of the basic SFT fields with respect to spherical radii, equations with similar dynamic characteristics of (2.35), (2.36a), and (2.36b) can be obtained. The main difference between SFT as it applies to the hydrogen atom is that instead of the cyclotron effect being in the r–θ plane, it occurs in the r–ϕ plane, the same as the orbital plane.

This gives a direct connection with the precession of the perihelion of Mercury. This phenomenon, so crucial in the historical validation of GR, is thus also connected to the SFT motions of Mercury. Coupled rotations of SFT both in the same plane are the origin of planetary precession including Earth's. We are not talking about the origin of the *angle* of the precession here which like a precessing gyroscope is 23.5° from the ecliptic due to the combined action of the Sun and the Moon on the Earth's motions, but Earth's ongoing energy of rotation once a day about its North–South axis. For instance, Copernicus estimated the precession of the equinoxes as 0.0127 degrees per year, while its modern estimate is 0.01396 degrees per year.

Polarization α is constant if the field is below a critical value at which the electron is stripped away from the proton. At this critical level, the structure of the matter will collapse to the next internal energy level. For example, inside solar masses where internal pressures build up, proton–proton interactions occur. The ratio $\frac{\alpha}{4\pi\varepsilon_0}$ is often given in tables of experimental data where R is assumed to be 10^{-10} m, and the polarization ratio is in units of 10^{-30} m^3. For example, for He $\frac{\alpha}{4\pi\varepsilon_0} = 0.201 \times 10^{-30}$ m^3. A force exists on the atom due to the external field acting upon the dipole moment formed by the two monopoles within the atom

$$\vec{F}_E = \frac{\vec{p}_E}{3} E = \frac{-q^2}{4\pi\varepsilon_0} \frac{\vec{d}}{R^3} \tag{2.37}$$

The fields at a point R due to an electric dipole p_E can be written in terms of the components along the radial and the axis of the dipole moment. The radial term $\vec{p}_E(r) \bullet \hat{r}$ is the component we are interested in where

$$\vec{E}_{dip} = \frac{1}{4\pi\varepsilon_0} \frac{3\hat{r}(\vec{p}_E \bullet \hat{r}) - \vec{p}_E}{R^3} \tag{2.38}$$

Using (2.38) a force exists between two dipoles

$$\vec{F}_{dip} = \vec{p}_{E1} \bullet \frac{\partial E_{dip}}{\partial R} = \vec{p}_{E1} \bullet \vec{p}_{E2} \frac{-3.2}{4\pi\varepsilon_0 R^4} \tag{2.39}$$

The difference to Coulomb's inverse-square law (2.36b) is obvious. If the force is to be an inverse square, this implies a particular form of dipole moment. The inverse-square form results if each dipole moment is directly proportional to R. Where conglomerates are involved, we integrate over all atomic moments within each, defining polarization volumes that include all matter within the two interacting objects. Portis discussed polarizability of diatomic alkali halide molecules showing that the potential energy function of the system of particles can be written as a sum of terms such as terms for the Coulomb interactions between the two ions, and a number of other terms, including a term $\frac{-1}{4\pi\varepsilon_0} \frac{2p_1p_2}{R^3}$ that represents the dipole interaction between ions involving the long-range inter-nuclear distance R. Each moment is a complicated function $p = qR\left(1 - \frac{\beta(\alpha_1+\alpha_2)+4\beta^2\alpha_1\alpha_2}{1-4\beta^2\alpha_1\alpha_2}\right)$ demonstrating that the force can be a complicated function but directly proportional to the separation. We can use a particular form of dipole moment to model gravity by an inverse-square law as in (2.36a). We assume the following form for the long-range dipole–dipole polarization.

$$\alpha_{dip1} = \frac{\vec{p}_{E1}}{\vec{E}} = \frac{4\pi\varepsilon_0}{3} d_1^3 R \tag{2.40}$$

All atoms in volume V_1 relating to conglomerate dipole p_1 act to provide an external electric force for the second dipole moment defined in terms of the polarization volume V_2, and vice versa. Assuming for simplicity that all moments at either end are separated by distance R then the integrations simplify and we obtain the dipole moments in the form

$$p_{E1} p_{E2} = qN_1 \frac{4\pi\varepsilon_0}{3} d_1^3 R_q N_2 \frac{4\pi\varepsilon_0}{3} d_2^3 R \tag{2.41}$$

where N_1 and N_2 are the numbers of atoms or molecules comprising dipoles 1 and 2. Importantly then α_{dip} is now directly proportional to the radial separation vector $\alpha_{dip} = \alpha_{dip}(R)$. The force F_{dip} given in (2.39) is now seen to be an inverse-square function of separation $F_{dip}\left(R^{-2}\right)$ as the in Newtonian formula (2.36a). The dipolar character of the matter comprising material of dipoles 1 and 2 dictates the particular form of interaction. Equation (2.40) has been assumed to match the experimental observation of Newton's inverse-square law.

At this point we rewrite the long-range dipole–dipole force equation that was developed theoretically in the previous section specialized for the Earth–Sun dipole–dipole forces.

$$F_{dip} = p_S p_E \frac{24\pi\varepsilon_0}{9R^4} = N_s d_s^3 R \varepsilon_{rS} N_E d_E^3 R \varepsilon_{rE} \frac{24\pi\varepsilon_0}{9R^4} \tag{2.42}$$

In (2.42), we have twice included a factor 3 from the volumetric definition of long-range dipole–dipole polarization (see footnote i) and also two relative dielectric constants (scalars) for the Earth ε_{rE}, and the Sun ε_{rS}. Examination reveals three basis groupings. First there is the gravitational constant G that we might think of as relating to neither Sun nor Earth but more related to both objects, or something that they somehow share; this is obviously a fundamental physical constant. Second, there are "body variables," namely the masses m_S and m_E. Finally, there is the inverse-square relationship of separation $1/R^2$. Thus we can organize the variables and

Table 2.1. Properties of Sun and Earth

	ρ (kg/m^3)	Radius (m)	Volume (m^3)	Mass (kg)	Atom size (m)	Dist R_S (m)
Sun	1.41×10^3	6.95×10^8	1.41×10^{27}	1.98×10^{30}	0.50×10^{-10}	
Earth	5.52×10^3	6.38×10^6	1.09×10^3	6.00×10^{24}	0.25×10^{-10}	1.49×10^3

constants into similar groupings to (2.36a). All numerical constants and the factor involving the permittivity of free space ε_0 can be formed into a *gravitational constants* group. Similarly, we group variables involved in the dipole moments into "mass variables," and finally we have the familiar inverse-square relationship. Calculations reveal a reasonable resemblance between (2.36a) and (2.42). The estimated gravitational constant is written

$$G' = \frac{24\pi\varepsilon_0}{9} \tag{2.43}$$

From this we find $G = 7.4 \times 10^{-11}$ Nm^2/kg^2 a discrepancy of about 11% from the Newtonian constant. Given this simple model only considers a two-body Solar system the accuracy is reasonable. The dielectric constants of our model Earth and Sun can be estimated from the numbers of atoms in the Sun N_S and Earth N_E. From the atomic volume of the Sun and from its mass and density (Table 2.1), we can estimate $N_S = 2.68\times 10^{57}$. As for the Sun, we obtain an atomic volume for the Earth's matter and from mass and density, we estimate $N_E = 8.31\times 10^{51}$. Forms of the dipole moments for the Sun p'_S and the Earth p'_E are written as

$$p'_S = qN_s d_s^3 \varepsilon_{rS} \tag{2.44a}$$

$$p'_E = qN_E d_E^3 \varepsilon_{rE} \tag{2.44b}$$

Noting that an atomic volume will occupy an electron and a proton we can convert the dipole factors (2.44) to masses. Using the known masses of the Sun and the Earth, we calculate the dielectric constants as the ratios of the masses to these dipole moment factors. Relative dielectric constant is a dimensionless quantity so mass in this system is equivalent to a sum of atomic volumes times the elementary charge times the relative permittivity of each planet. We calculate $\varepsilon_{rS} \approx 5.9$ and $\varepsilon_{rE} \approx 4.6 \times 10^6$.

Table 2.2. Composition of Sun

Element	% of mass
Hydrogen	71.0
Helium	27.1
Oxygen	0.97
Carbon	0.40
Nitrogen	0.096
Silicon	0.099
Magnesium	0.076
Neon	0.058
Iron	0.014
Sulphur	0.040

Both Sun and Earth will have a mean relative dielectric constant that includes non-metallic elements near the surface and heavy metals at the solid core. The dielectric model of the Sun does appear to be reasonable given its estimated composition given in Table 2.2. For the Earth, the core is known to be predominantly iron and nickel. There is a solid inner core and a liquid outer core. The mantle, Earth's biggest layer extending to a depth of 2890 km, consists of silicates moving slowly under thermal convection. These silicates are relatively rich in iron and magnesium. The crust is less metallic and more heterogeneous than the inner layers. With regard to our dielectric model, Earth is a very good conductor surrounded by silicates. Hence Earth is a good model for a DEM ideal for acting as a gravitational body holding water and oxygen in abundance within its surface.

Over the past 50 years a theory of intercontinental plate tectonics has emerged that describes the very slow movement of plates (10–40 mm per annum) within the Earth's lithosphere. Various directions of the magnetic field were found in rocks of differing ages; spreading of the sea floor was associated with magnetic reversals. The theory grew that plates move by the release and redistribution of heat energy from Earth's interior. On the other hand the SFT dipolar model for gravitation suggests that all matter on Earth is subject to rotations due to the dipole–dipole interactions of the atomic magnetic moments similar to the dielectric forces. Since the composition of Earth is basically such that heavier elements sink towards the centre at a faster rate than lighter elements due to these Newtonian forces, there will a magnetic mixing due to Earth's inhomogeneity and this will be greater nearer the centre due to the presence of crystalline heavy metals. Indeed the Earth's core was recently discovered to be rotating faster than the surface by an extra turn per 120 years, very much faster than continental drift. At the core the B-fields are considered about 200 times those at the Earth's surface. So the rate of rotation at the core is around 20,000 that of continental drift, approximately the square of the relative B-fields.

The Earth's rotation may have a similar genesis to the advance in Mercury's perihelion, i.e. the rotation is due to the magnetic torque which Newton's law does not incorporate but Einstein's GR does. Due to Earth's inhomogeneity across depth, latitude, and longitude magnetic mixing is occurring at all times, at all depths, and includes an atmospheric mixing that varies according to the time of the day and season of the year due to the Moon and the Earth's axial tilt. The variation in atomic magnetic moment may be underlying cause of both continental drift within the lithosphere and earthquakes at the Earth's surface. There may well be relationships with the motions of the Moon, planets, and the Sun that cause peaks in Earth's geomagnetic field strengths. Researchers have been attempting with some success to predict earthquake occurrence by planetary configuration.

At the same time, the intense pressures at the centre of the Earth may be capable of causing thermonuclear reactions to take place hence radioactivity is an ongoing process and not a half-life process of decaying radioactive matter. According to SFT, nuclear material is involved in a dynamic balance at the galactic level. The black hole near the centre of our Galaxy has suns and planets in balance with it that are capable of exchanging bosons in a similar way to the electron and proton within the hydrogen atom and planets and Sun of the solar system. In this case the mediator at the galactic level is probably gluonic due to the solid-like rotational velocity observed in galaxies. Magnetic reversals of the Sun and Earth may be related to the photon's composite structure; the photon has a seemingly random spin related to the energy density of its surroundings. Hence SFT adds novel perspectives to a number of scientific observations.

In terms of gravitation, SFT may add new levels of detail to the theories of Newton and Einstein, thus extending the horizon of our knowledge.

The preceding analysis of the long-range dipole–dipole forces of atoms may result in a resonance phenomenon at planetary domains although this needs to be averaged over time, space, and all participating atoms. In other words equilibrium may not hold and planets are for most of their lives in slow evolution of their orbits and spins. Their motion is however always a geodesic motion when averaged. Within these gravitational forces, the overall internuclear distances between any two atoms may not always be an integer multiple of $\pi/2$. On the other hand examining the hydrogen molecule as two hydrogen atoms being brought somehow close together, we find that there are radial positions of resonance where such an integer multiple holds and the dipole–dipole forces are minimized. Depending on the atomic frequencies the integer multiple positions may be resonance positions. Upon reaching their minimum state of separation the two atoms may bond together to form a hydrogen molecule. Pauling referred to this resonance within chemical bonds. Within the long-range dipole-dipole forces the inter-nuclear distances are averaged over all the remnant non-resonant forces, forces left over beyond the resonant positions that are tugging to and fro, repelling and attracting to be averaged out over all atoms and molecules as the force we know as gravity. Hence there are specific frequencies emitted by atoms and molecules within the Sun that relate to atoms within the Earth. This creates a signature of frequencies that help us analyse the Sun using spectrographic data as shown in Table 2.2.

In the hydrogen atom the binding energy, the photon, transits between proton, and electron in the r–ϕ plane, collides elastically with the electron thus changing its plane to the r–θ plane then transits back to the proton where it collides elastically with the proton thus finally changing it plane back to the original r–ϕ plane. This cycle is repeated over and over with a very precise photon stream in the two planes of its motion r–ϕ and r–θ. In fact there are two such streams, one going from proton to electron, the other going from electron to proton. In the case of the SFT of the long-range dipole–dipole forces similar streams of photons move between the

Sun and Earth and back to the Sun. What we should expect on average is that there will be a relatively narrow beam of photons emerge from the Sun in the r–ϕ plane of the Solar System spreading out to cover the entire equatorial region of the Earth. This beam will change direction to emerge from one of the Earth's poles and move in the r–θ plane back to the Sun where it enters the Sun's pole. There will be two such streams of photons one going from Sun to Earth, and the other going from Earth to Sun. Physically what we may find is that these streams exist in deep space as *interplanetary superhighways* of energy waiting to be discovered by interplanetary probes. If a probe should enter one of these streams they may well find the energy density rises beyond that of ordinary space where no such streams exist. In fact a similar system of streams has been used by Lo working at the Jet Propulsion Laboratory at California Institute of Technology and the NASA to design the Genesis mission flight path. The streams have been described as "a 'freeway' through the solar system resembling a vast array of virtual winding tunnels and conduits around the Sun and planets... can slash the amount of fuel needed for future space missions." At present such streams are applied as staging points for interplanetary travel. Similar streams composed of various bosons may exist at all levels of gravitational structure within the Universe (see Chapter 3).

2.14 Strong Nuclear Fields

One important aspect of SFT and the bi-spinors that represent the EM motions of the electron, the proton, and the photon, is the possibility that there is an associated relationship based on trispinors that represent the strong nuclear motions of quarks inside the proton. As is well known, there are three quarks inside the proton. In this case there is a self-field interaction between triplets of quarks where the interaction sub-matrix is of size 3×6 per particle as distinct from the EM field sub-matrix that is of size 2×4. Now each particle has *three* spinors that mutually link the three quarks together. Each spinor operates in a plane and the three planes form an orthogonal set in 3-D space. This means that there exists a system of equations similar to Maxwell's equations but having additional differential components such as to form four scalar equations instead of the three associated with Maxwell's EM equations.

Fleming and Bauer (nee Colorio) have recently suggested that the strong nuclear fields controlling the motions of charged particles satisfy the following adaptation of the ML equations. In general, the region is assumed isotropic and homogeneous and ε_n, μ_n and μ_n are invariant scalars. Where nuclear sub-particles, quarks, carrying units of elementary charge q_q are studied, the modified Maxwell-Lorentz equations can be written as

$$\nabla \bullet \vec{E} = \frac{q_q}{v_q} \tag{2.45a}$$

$$\nabla \bullet \vec{H} = 0 \tag{2.45b}$$

$$\nabla \bullet \vec{N} = 0 \tag{2.45c}$$

$$\nabla \times \vec{E} + \mu_n \frac{\partial \vec{H}}{dt} = 0 \tag{2.45d}$$

$$\nabla \times \vec{H} - \varepsilon_n \frac{\partial \vec{E}}{dt} = \frac{\pi}{s_q} q_q \vec{v} \tag{2.45e}$$

$$\nabla \times \vec{N} + \nu_n \frac{\partial \vec{E}}{dt} = 0 \tag{2.45f}$$

where the modified Lorentz equation for the forces acting on the quarks is

$$\vec{F} = q_q \vec{E} + q_q \vec{v} \times \vec{B} + q_q \vec{v} \times \vec{M} \tag{2.45g}$$

and constitutive equations $\vec{B} = \mu_n \vec{H}$, $\vec{D} = \varepsilon_n \vec{E}$, and $\vec{M} = \nu n \vec{N}$, where ε_n, μ_n, and ε_n are invariant scalars, the nuclear constitutive parameters similar to those of free space, ε_0 and μ_0, except the energy density within the nucleus now depends upon the three gluon fields $dU_N = \rho_N dV = \frac{1}{2}(\varepsilon_0 \tilde{E} \bullet \tilde{E} + \mu_0 \tilde{H} \bullet \tilde{H} + \nu_0 \tilde{N} \bullet \tilde{N}) dV$, where $\tilde{N}$ is termed the nuclear field and $\tilde{M}$ is the nuclear flux density. As with $c = (\varepsilon_0 \mu_0)^{-1/2}$, there are corresponding relationships between the gluon speed and the ratios of the three fields. These modified equations provide three orthogonal motions per quark. There are 6 unknowns per particle and 18 in the nuclear system.

The curl equations (2.45d–f) provide four scalar equations, and there are two virial equations to give six equations with six unknowns. Like the photon, the resulting analytic parametric solutions may be compared to the experimental results given by particle physics.

In the strong nuclear case there are 3×4 scalar curl equations plus 3×2 force balance equations making 18 equations in 3 particles altogether that can be solved to yield 6 quantum numbers per particle in nuclear structures. This compares with the 2×3 scalar equations from Maxwell's curl equations plus 2 force balance equations totalling 8 equations per 2 particles, giving 4 quantum numbers per particle for atomic structures. The two extra quantum numbers agree with the experimental observations of high-energy physics. The 3 quarks have 3-way streams so that each particle performs spinor motions in 3 planes and not 2 as in the case of EM related particles such as the proton and the electron. Now of course the only way that the quarks can perform these additional spinor motions is if the photons themselves have formed into compounds *beyond* "ordinary" or low-energy photons. When photons are inside high-energy regions they may exist in higher energy states and compounds. Like atomic structures then, photons may *also* form into structured compounds. In the case of quarks, the photon particles may have a third spinor motion when they form into triplets. The forces these photonic compounds carry yield the third component of spatial direction and they *only* carry this extra force component *inside* the nucleus.

2.15 Photonic States and Compounds

The search for an upper limit of the photon's rest mass has continued over many decades. As discussed by Jackson one aim of the search is to quantify the cut-off between classical electrodynamics and QED and to establish the accuracy of Coulomb's electrostatic square law. Different experiments have tightened the limit downwards. Due to Earth's geomagnetic field, fluctuations in the torque of a Cavendish balance are observed. The Jovian geomagnetic effect has similarly been measured. In addition, magnetic fields have been found to

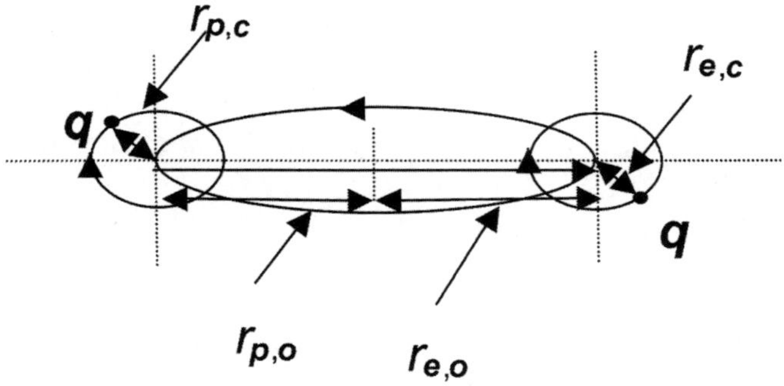

Figure 2.14. Composite photon: SFT provides a rationale for the differences between (a) quantum physics having a discrete energy spectrum as its sub-particles masses are unequal; only certain radii and velocities are valid solutions to the SFT equations; and (b) continuous physics having a continuous energy spectrum as its sub-particle masses are equal, all radii and associated velocities are valid solutions to the SFT equations.

permeate the cosmos arising if photons possessed mass during the big bang. The current experimental best estimate is $m_{ph} < 4 \times 10^{-51}$ kg. In Chapter 3, we determine a theoretical estimate of the mass of the photon. An internal dynamics can be associated with this photon mass where SFT provides a bi-spinorial solution for the photon. It is suspected from known high-energy physical data that photons have similar states and compound structures to atoms. The "ordinary" photon of free space may consist of two internal particles that rotate around each other in order to balance their dynamic energies, as in the hydrogen atom. Since the energy response of the photon is continuous, the two particles must have equal mass. The difference between quantum and continuous physics is thus seen as the ability of continuous systems to solve the underlying equations at any frequency due to the mass equality or symmetry of their sub-particles.

Thus photons are seen to obey SFT as do atoms (compare Figure 2.14 with Figure 2.7). Inside atoms there are central nuclear regions of enhanced energy densities where photonic states and compounds might exist due to the slower moving nucleus sweeping out larger volumes relative to the faster electron. Photon states and compounds may help explain some important physical and biological phenomena. Schumann resonances occur at extremely low frequencies (8, 14, 20, 26, 32 Hz, etc.) the result of the Earth ionosphere region having a discrete geometry, a series of spherical layers, that forms a large resonant cavity waveguide whose resonances are sensitive to variations in temperature. These layers may be due to the increase in energy density as the Earth is approached or departed by photons.

Ionospheric layers may also be layered into various forms of chemistry such as cloudless and cloudy layers according to the photon states involved. The phenomenon of snowflakes may be due to the myriad photon states involved in ice molecules surrounding dust particles etc that fall to Earth. Avalanches may be due to the seemingly random nature of the spin states of photons that change very abruptly with small changes in temperature or ambient energy density. All states are valid as long as the sub-photonic particles are of equal mass.

The energy states of individual biological cells and tissues composed of many cells acting in cooperation may depend on photonic mechanisms and on the dynamic balances given by SFT. First, H_2O in the cell may take liquid or crystallized multi-layered forms inside and around chromosomes and other macromolecules. Such structures may depend on *slow* changes induced by shifts between photonic states driven by changes in energy. Second, cell division may depend on *sudden* changes induced by similar photonic shifts. The E- and H-fields in and around DNA and other structures such as the extra cellular matrix may change abruptly over the cell cycle.

Dramatic cellular changes are observed: the chromosome first aligns at the cell equator in equilibrium between the spindle poles, it then cleaves into halves; each chromatid replicates into a complete chromosome that moves to its spindle pole; the membrane walls cleave at the equator into two complete cells. The energy distribution inside cells and tissues may reflect the type of dynamic balance seen within the hydrogen atom.

Chapter 3

The Photon

Light has played a major role in civilizations across human history. The concept of the photon, the quantum of EM energy, goes back to Newton and more recently to Einstein's photoelectric effect. The photon has remained enigmatic due to its dual wave–particle nature. Since quantum techniques were formulated around 1927, the photon has remained probabilistic due to Heisenberg's uncertainty principle (HUP) where velocity and position cannot both be known. If a composite photon, or equivalently a bi-spinorial EM field is assumed, HUP is obviated and the self-field theory (SFT) formulation results in deterministic internal photon motions. In fact HUP and SFT are theoretically almost identical in form. From 1923 to 1926 using a 100 inch telescope at Mount Wilson Observatory, Hubble discovered that the local group "stars" were in fact other galaxies outside our Milky Way. Similarly the discovery of structure within the photon is an extension of our physical knowledge at the very small end of the size spectrum. It appears there is an identifiable organization behind the structure of photons and other field particles, bosons, known to particle physics. Previously thought to be "elementary," the composite structure of ordinary photons appears similar to that of the hydrogen atom obeying a similar SFT system of equations. This provides a "photon chemistry," similar to atomic chemistry, to explain some features of bosons including W^+, W^- and Z^0 bosons of the electroweak force and gluons of

Self-Field Theory: A New Mathematical Description of Physics
By Tony Fleming

www.panstanford.com

the strong nuclear force. Phonons also sit under this theoretical umbrella and a stoichiometric equation can be written in terms of gluons, photons and phonons. Three gluons rotating in three orthogonal directions fit within a modified system of Maxwell–Lorentz equations providing a basis for a dynamic solution inside nuclear regions.

3.1 Introduction

Ancient civilizations across the globe noted relationships between cosmological events such as the summer and winter solstices and terrestrial life including agricultural, hunting, and gathering patterns. Incorporation into architecture of special times is found in prehistoric and ancient edifices. Stonehenge in southern England (Fig. 3.1), the pyramids at Giza in Egypt, and the acropolis in Alatri have all been studied as to alignment with astronomical sources of light. Alignment can be related to creation mythologies. Not only is ancient architecture related to astronomy and religious beliefs, the Torre dei Venti in the Vatican (Fig. 3.2) shows the Sun's path on its floor thus revealing the Gregorian calendar, the international standard for date based on the motions of the Sun, the Moon, and the presumed birth year of Jesus Christ. From prehistory through ancient times light has remained a central focus for civilization.

Figure 3.1. Stonehenge. (Source: http://www.english-heritage.org.uk/server/show/nav.16465/chosenImageId/12) See also Color Insert.

Figure 3.2. Torre dei Venti. (Source: http://asv.vatican.va/it/arch/torre.htm) See also Color Insert.

According to modern science, light consists of photons, quanta of EM energy, considered by many to be elementary since their experimental discovery by Einstein in 1905. The photon is involved across the widest range of physics from the cosmological big bang to the motions of the atom and the workings of the DNA. Applications of photon technology beginning before the industrial revolution to the present are numerous. The telescope, the microscope, residential lighting, radio, telecommunications, radar, computers, robotic, and mobile gadgetry are all part of everyday domestic, industrial, and scientific life today. Nanotechnology represents the immediate future of photonics, while biophotonics is ready to revolutionize medical practice allowing therapy via EM and/or vibrational appliances. The term photon, proposed by Lewis in 1926, was accepted by the physics community 25 years after Planck's discovery of the quantum nature of EM radiation. Since those days the photon has remained a theoretical blur, a point of uncertainty according to Heisenberg. To quantum field theory (QFT), it is a wave-packet, a form of Dirac delta function, a singularity in the fabric of space-time, as enigmatic to quantum mathematics as black holes are to modern gravitation.

The history of light as coherent particles performing dynamic motions goes back to Newton who considered the speed gradient of ordinary light across a material interface and the reasons why separation occurs into different colours via refraction through a prism. Being interested in Galileo's experiments using objects of different masses dropped from the Tower of Pisa, Newton considered the free fall of cannon balls due to a range of initial horizontal velocities. If the cannon ball had sufficient velocity it might eventually go into orbit around the Earth. He felt this was a sufficient analogue to consider light consisting of tiny corpuscles. Grimaldi had first recorded the diffraction of light via a single slit experiment. Young showed in a double slit experiment that light produced an interference pattern due to diffraction. Huygens and Fresnel showed that this phenomenon validated the wave theory of light. Fraunhofer recorded hundreds of solar spectral lines that would later be understood by Bohr's theory of atomic emissions. Coulomb, Ampere, Gauss, Biot, Savart, Faraday, and Heaviside all contributed to the eventual theory of EM as coupled partial differential equations by Maxwell. Incorporated into the equations, similar in form to Newton's gravitational law, was Coulomb's inverse-square dependence between electric field and radial distance. Like magnetostatics, this was considered an uncoupled form of EM field. On selecting a gauge associated with the particular symmetry of a chosen EM field, Maxwell's first-order equations could be written as second-order potential wave equations. Considered by some to simplify the solution, the scalar and vector potential wave equations revealed light to be a wave, a particular group in an infinite spectrum. Of all these wave solutions only a certain group having gauge invariance obeyed Maxwell's first-order equations. Gauge was known to classical calculus via conditions holding between first and second-order equations an inverse form of integration constant.

Planck's self-termed "act of desperation" succeeded in discovering the quantum basis of black body radiation including the quantum of energy h. Others also investigated this quantized EM energy. Einstein investigated this via the photoelectric effect in metals and Compton via scattering of X-rays by electrons. The wave–particle duality of light was now an experimental fact, if not yet accepted theoretically. Einstein's discovery of the special and general relativistic nature of light was the culmination of efforts to understand why Maxwell's equations were invariant to coordinate transformations named after Lorentz who discovered them. Newton's inverse-square law of gravitation was modified by relativity. Poincaré had set up a conceptual framework for Einstein to follow while Michelson and Morley performed an experiment giving evidence that the speed of light did not depend on direction. Many since have used this to infer there is no ether as the Earth's motion should have produced a measurable difference in speed with change in direction. That relativity is associated with the internal motions of the photon also provides reason for the results. In special theory of relativity, Einstein assumed the equations of physics are the same in all inertial frames; and the speed of light c is the same in any inertial frame regardless of whether the light is emitted by a body at rest or in uniform motion. From these postulates he derived the Lorentz transformations, and the Fitzgerald–Lorentz time and space dilation. Einstein showed any ether to be superfluous to his deductions.

From their origins early in the 20th century, both quantum mechanics and relativity forged wide theoretical and technological paths across science. Developments led to quantum electrodynamics (QED) and quantum chromodynamics (QCD), and the standard model of particle physics. Amidst the theoretical and technological triumphs grew a conviction based on gauge symmetry that the photon mass must be precisely zero. Perhaps this conviction had its origins in the fact that the most important application, atomic weaponry used on Hiroshima and Nagasaki, concerned huge amounts of energy. In comparison, the self-energy of the photon is insignificant. Yet in retrospect this was tantamount to throwing out the baby with the bathwater. Within the notion of a photon with mass sits an entire physics waiting to be understood complete with its own spectroscopy and resonant frequencies. The photon, far from being a blur, like the "stars" that Hubble found to be galaxies, has a structure waiting to be uncovered.

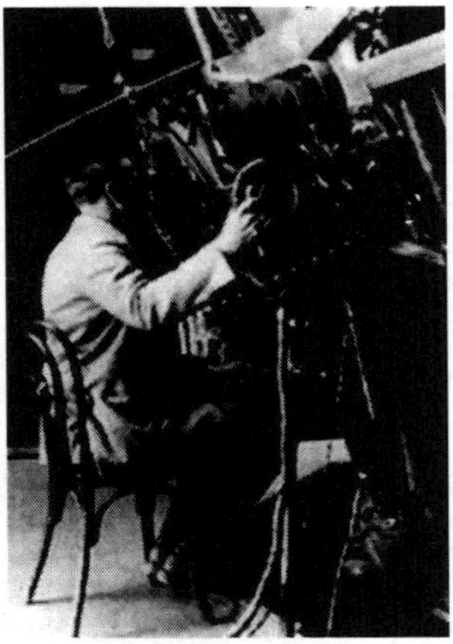

Figure 3.3. Hubble at the 100 inch telescope at Mount Wilson. (Source: http://www.soteoria.hpg.ig.com.br/Hubble/page1.htm)

Many advocates of a zero-mass photon base their opinions on either a range of cosmological evidence relating to Hubble's observed cosmological redshift (Fig. 3.3), on the postulates of relativity, or alternatively the potential basis of QFT. One issue is termed "tired light" where the photon loses energy with distance; in its simplest form this has the form $E(r) = E(0)e^{-r/R}$. The accepted theory of redshift is that the Universe is currently expanding subsequent to an initial big bang. The motion of photons in deep space along giant cosmic circles shows how massive photons can move in keeping with self-field theory (SFT) and the constraints of relativity without losing energy. But this says nothing directly about its motion while radiating from mass to mass as in for instance the binding energy within the atom, only how it can move as a tiny mass within the cosmos on a giant circle in dynamic balance with the centre of mass of the Universe. Does the photon lose energy as it transits between proton and electron, and if other universes exist does the photon lose energy in binding our Universe to other universes in a kind of giant crystal lattice? And what sort of forces might be involved? It may be that Bose–Einstein condensates form at the near-zero temperatures in this part of deep space. If photon mass exists what does it say about the size of the universe? It may be that redshift is both a result of expansion following the big bang *and* tired light as the photon and other bosons radiate outwards to and inwards from other universes. Up to the present our understanding has limited relativity to the *external* motions of the photon. SFT on the other hand provides a view where relativity consists of both internal and external motions of photon. This more physical understanding of relativity gives insights of a non-zero-mass photon and its role in our Universe and the cosmos generally.

Lambda cold dark matter (ΛCDM) is the standard model of big bang cosmology. According to considered opinion "ΛCDM attempts to explain both the existence and structure of the cosmic microwave background, the large scale structure of galaxy clusters and the distribution of light elements, and the accelerating expansion of the universe observed in the light from distant galaxies and supernovae. It is the simplest model that is in general agreement with observed phenomena." ΛCDM is based on the cosmological principle in which our Universe is assumed homogenous and isotropic. The Cosmic Background Explorer (COBE) satellite discovered in 1991 that the temperature of the background radiation varies by less than 1 part in 100,000 over any two directions in the sky. However, in 2008 researchers determined that all clusters out to 6 billion light-years are moving about 0.9×10^6 m/s towards a 20° region near the constellations of Centaurus and Vela.

According to the cosmological principle the constitutive parameters of deep space $\varepsilon_0(r)$ and $\mu_0(r)$ are also assumed constant throughout the Universe. Looking inside the hydrogen atom such a model for the energy density does not hold. The atom appears spherically variable in its energy density; as the nucleus is approached from outside, $\varepsilon_A(r)$ and $\mu_A(r)$ increase to energy levels that allow the weak and strong nuclear bosons to exist. Any increase in $\varepsilon_A(r)$ and $\mu_0(r)$ may be inverse functions of position, e.g. $1/r^2$. Similarly, it may be that following the big bang, there is an inhomogeneity moving outwards from its initial location. The speed of light across the Universe may depend on constitutive parameters as $c = [\varepsilon_U(r)\mu_U(r)]^{-1/2}$ where $\varepsilon_U(r)$ and $\mu_U(r)$ are functions of radial position within the Universe where r is taken from the position of the big bang. It appears at first sight unreasonable that the universe is inflationary *and* isotropic and homogeneous at the same time. Watching the progression of an atomic blast reveals the type of process that might have occurred at the big bang. The redshift may in part be due to this inhomogeneity with the speed of light increasing as the energy density decreases. However, due to the immense size of the Universe it could be that the solid angle of any resulting dip in the redshift due to the centre of mass or its edges are either tiny or smeared out. Perhaps ultra-low-noise-level scans of cosmic microwave background (CMB) radiation are required to define the degree of inhomogeneity in more detail. Whatever the case, the evidence is strong that redshift depends on the big bang and the evidence of CMB gives the age of the Universe as 13.7 billion years.

In terms of the photon's potential theory, Proca put forward a Lagrangian action for a massive field. In the 1920s he worked at the Marie Curie's *Institut du Radium*. She supported his work that was strongly connected to the weak nuclear phenomena she and her husband had studied since 1895. Proca formulated the following action describing a massive spin-1 field of mass m in Minkowski space–time.

$$L = -\frac{1}{16\pi}(\partial^{\mu} A^{\nu} - \partial^{\nu} A^{\mu})(\partial_{\mu} A_{\nu} - \partial_{\nu} A_{\mu}) + \frac{m^2 c^2}{8\pi \hbar^2} A^{\nu} A_{\nu} \quad (3.1)$$

This gives rise to a modified set of Maxwell equations called the Maxwell–Proca equations where the mass of the photon is non-zero.

$$\nabla \bullet \vec{E} = \frac{\rho}{\varepsilon_0} \quad (3.2a)$$

$$\nabla \bullet \vec{H} = 0 \quad (3.2b)$$

$$\nabla \times \vec{E} + \mu_0 \frac{\partial \vec{H}}{dt} = 0 \quad (3.2c)$$

$$\nabla \times \vec{H} - \varepsilon_0 \frac{\partial \vec{E}}{dt} + \frac{\lambda_r^2}{\mu_0} \vec{A} = \vec{j} \quad (3.2d)$$

The photon mass is given in terms of the Compton wavelength

$$m_y = \frac{h\lambda_y}{c} \quad (3.2e)$$

These equations include a magnetic vector potential as well as the E- and H-fields. Normally discussion of a non-zero-mass photon concerns only a single photon with the Proca–Lagrangian or

Maxwell–Proca formulations, not the mutual fields, the two photons of SFT. Similarly the original SFT of Abraham and Lorentz was applied to a single particle, the electron, and its isolated self-field. As it stands the Proca action concerns one photon. It is not until *two* photons are examined that the action of equation (3.1) can be arranged to be constant. In the same way, one uncoupled charged particle cannot be dynamically stable in a local sense. As with the electron and proton, neither can be confined nor stable without the other. Their stable centre of mass can only be found by taking *two* photons, by examining *both* E- and H-fields together. Indeed it was the dynamics due to the H-field that was lacking from Bohr's original theory.

Theorists in the late 20th century used an interplay between potential theory, gauge symmetry, and experimental observation in an attempt to unify our understanding of the large number of particles known to particle physics. The overall idea arose that since scalar and vector potentials were involved in EM, then an extension to tensor potentials might be a valid step forward. To start, using the example of the EM fields, Glashow, Weinberg, and Salam in 1968 derived an electroweak theory for the electromagnetic and weak nuclear forces. As well as predicting a zero-mass photon they predicted the massive W and Z bosons that were found in 1981 with the correctly predicted masses. By extending this electroweak theory to a tensor form the standard model of elementary particles (Fig. 3.4) attempts to unify the EM, weak and strong nuclear

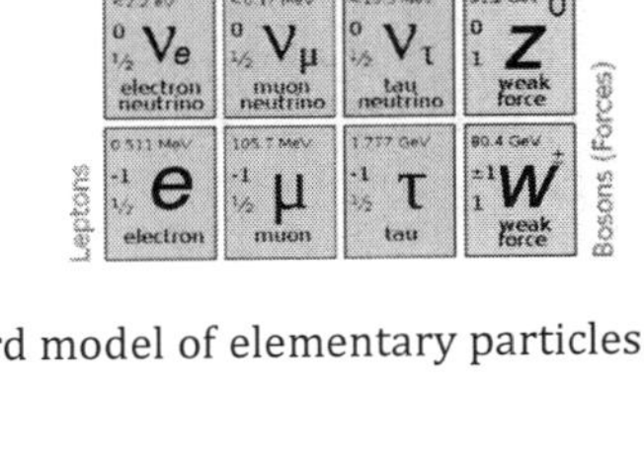

Figure 3.4. Standard model of elementary particles. (Credit: Wikimedia.) See also Color Insert.

interactions and the particles involved in them. A gauge symmetry theory in the gauge group SU(3)×SU(2)×U(1), the standard model does not include gravity, dark matter, nor at present does it describe non-zero photon and neutrino masses. The strong interaction was validated in 1973–1974 when hadrons were found to be composed of quarks. The use of gauge groups has attempted to find groups that correspond to the observed behaviour of the various interactions, including the confinement of the strong nuclear interaction. The complexity of this form of mathematics precludes a detailed discussion within this text. Suffice to say some theorists take the existence of a non-zero photon as anathema to the potential theory built up around QFTs including QCD. As we have seen, however, non-zero-mass photons exist in atoms in pairs and as long as they have a fixed phase relationship between them, symmetry is maintained and is *not* broken.

Why have boson pairs or triplets not been examined by particle theorists? Basically the experiments focus in on one particle at a time. In a collider one particle is considered the target. In QFT the Lagrangian equations concern only one particle, its mass, and its field as was the case in the early studies of the electron's self-field. In QFT if the field has mass, gauge invariance is destroyed. On the other hand SFT can indicate when a system of particles is becoming locally unstable as well as determining dynamic equilibria. In atomic terms SFT sees levels of stability where a photon for instance is attached either to an atom or to some other physical system such as the plasma within a spectrometer; in both cases a dynamic equilibrium takes place, the photon is stable in the initial and the final states. Instability is the small period when a particle is actually changing its equilibrium state, like the lines on a phase diagram. SFT promises to reinvigorate the theory of particle physics by allowing a sub-photonic structured view of bosons. Such a view termed *photon chemistry* is similar to the atomic chemistry known since Mendeleev in 1869 discovered the periodic table by arraying the elements in a two-dimensional format according to their similarities in properties. According to photon chemistry the various bosons are modelled as composite charged particle structures in a similar fashion to atoms where each structure can be validated according to SFT starting with the photon and working upwards by changing the energy densities.

3.2 Analytic Estimate for the Mass of the Photon

Since the early days of QFT, the years leading up to World War II when QED evolved, symmetry has been connected with the various Lagrangian formulations. Similarly real eigenvalues require symmetry or skew symmetry of the interaction matrices resulting from the numerical implementation of the particular QFT. At the same time gauge calibration of the metric $g \to e^{2\lambda}g$ produces a transformation $A \to A - d\lambda$ of the potentials. Thus the photon, the quantum of EM energy, has been assumed to be massless. During this same period, no other option for the mathematical description of the electron within the atom was available. For the QFTs, gauge invariance is also assumed to imply a massless photon, likewise for other gauge bosons, otherwise spontaneous symmetry breaking would be observed. Jackson discussing the experimental limit on the inverse-square law of electrostatics concluded that the photon mass can be assumed to be zero throughout the classical range of distances and deep into the quantum domain as well. The Particle Data Group lists the best estimates on the upper limit of the photon mass giving the current limit as less than 2×10^{-16} eV by Lakes who measured the torque on a toroid Cavendish balance and obtained the limit via the Maxwell–Proca equations.

Recently experimental and theoretical findings have arisen to challenge the zero mass assumption. The neutrino, a member of the lepton family of particles, was found to have a finite, non-zero mass. The SuperKamiokande experiment in Japan put the mass at about 0.1 eV compared with the rest mass of the electron 0.511 MeV. In the experiment muon and tau neutrinos resulted in asymmetric oscillations that depend on the existence of differing non-zero neutrino masses.

Theoretically the self-fields of a *single* particle were studied in 1903–1904 by Abraham and Lorentz. They modelled the electron as a charged spherical surface of finite radius and found inconsistencies with classical EM theory as the radius approached zero. In 2005 the self-fields of *pairs* of particles were finally understood as a mutual phenomenon. The singularity problem at the charge points, similar to that afflicting the QFTs, was resolved by using periodic motions that avoid the charge points by their motions. The assumption is made that at equilibrium the two particles rotate never residing at their own centres of rotation. The SFT model for the EM field as a moving stream of photons provides an analysis of the hydrogen atom and yields an analytic derivation of Planck's number $\hbar = h/2\pi$. The bi-spinorial function for each particle provides a physically plausible interpretation of relativity. The "beads on a string" stream-like EM field modifies the field laws at the atomic level that must now be measured between two centres of rotation rather than between charge points. The atomic self-field motions are obtained as closed-form expression using the Maxwell–Lorentz (ML) equations. Depending on permittivity, permeability, energy density, spin of the particles, particle–mass ratio, and composition of the bosons, SFT has application across a wide range of physics domains from the photon upwards. SFT involves photon chemistry that gives insight into the internal structure of fields and particles. As we shall investigate, there is a theoretical relationship between the photon and the electron–positron pair that supports the experimental findings of annihilation and creation of particle pairs.

In 2007 SFT was used to study the binding energy of the hydrogen atom. A single period of the electron and the proton inside the hydrogen atom was investigated. This revealed the presence of two new quantum numbers linked to the photon states within the atom. This photonic mechanism promises to extend current predictive mathematics within chemical interactions such as the way water binds to molecules within both the weather cycle and the cell cycle. As shown below, the electron–proton transit can be used to obtain an analytic estimate of the photon mass by balancing Planck's energy with the energy of the photon within each atomic cycle. Gauge is maintained if the two photons transiting between electron and proton are of conjugate spin. In SFT terms the system is stable and periodic if a synchronous double rotation is maintained by the E- and H-fields where the orbital and cyclotron separations between electron and proton both differ in phase by π. On their own such synchronous orbits are unstable, but together they form stable dynamic equilibria. Using a modified form of the six ML equations, similar dynamic equilibria may also exist for strong nuclear regions

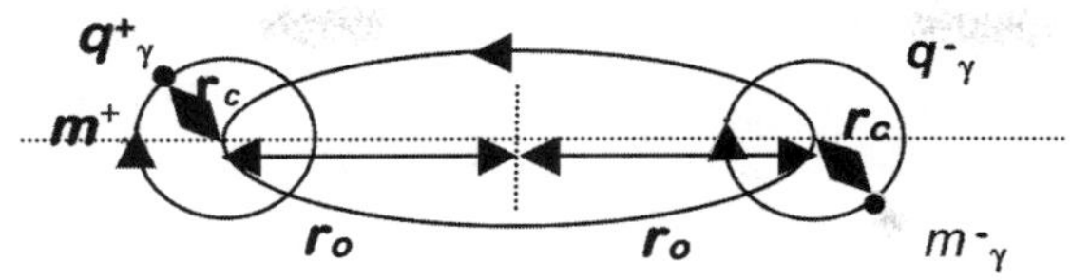

Figure 3.5. Photon modelled as two particles, the ephectron and the phroton, of equal masses.

involving triple rotations where gluons are known to have mass and the quark motions differ in phase by $2\pi/3$.

If the intrinsic energy of the system changes from that of free space ε_0 and μ_0 the EM fields within an atom can change thus altering the atomic and molecular binding structures. In effect this means that the radial and spin states of the two photons, one for the E- the other for the H-field involved in the binding energy can adapt in equal fashion to each other to the energy change where the photons themselves are assumed to have a composite structure. The EM ML system of two curl and two div equation has six variables including the E- and H-fields in response to the ambient energy conditions. The two extra variables give a range of variation orthogonal to the phase diagrams of atoms. A typical phase diagram shows only a solid line separating the various phases into solid, liquid, or gas indicating a very narrow range of energies between these common phases. SFT indicates a small range of binding structures that depend on the intrinsic energy of the system.

In a direct analogy to the SFT model of the hydrogen atom, one possible structure of the isolated photon is hydrogenic as shown in Fig. 3.5; two sub-photonic particles termed the ephectron and the phroton have equal mass and opposite charge. The equality of masses provides for Einstein's photoelectric energy $E = h\nu$ and its continuous energy–frequency response unlike the discrete physics of the atom where the electron and the proton have differing masses. The sub-photonic E- and H-fields for the two interacting particles are calculated using a SFT formulation similar to that applied to the hydrogen atom in Chapter 2. The primary parameter in the analytic spectroscopy of the ordinary photon is its mass, similar to the principal mode of the hydrogen atom where the electron mass specifies the spectroscopy. If the photon mass is known, the spectroscopy of photons can in principle be examined. The transition frequencies are expressed in terms of a continuous series

$$\nu_\gamma = R_\gamma \left(\frac{1}{m^2} - \frac{1}{n^2} \right) \tag{3.3}$$

where $m = 1, 2, 3, \ldots$ and $n = 2, 3, 4, \ldots$ The photon Rydberg number

$$R_\gamma = \frac{q_\gamma^4 m_\gamma}{8\varepsilon_0^2 h_\gamma^3 c_\gamma} = \frac{m_\gamma 8^2 \pi^3 \varepsilon_\gamma v_i^2 \alpha_\gamma}{q^2} \tag{3.4}$$

is a photon-specific function of mass and charge where we apply the corresponding expression obtained from the hydrogen atom in equation (2.29). Each photon-specific term m_γ, c_γ, α_γ, and q_γ, corresponds to terms in (2.29); c_γ is the speed of the sub-photonic field. A priori, the charge on the ephectron and the phroton is unknown and the permittivity ε_γ and the permeability μ_γ is particle specific.

The photon is assumed to exist in an ambient energy bath and to remain in a dynamic equilibrium like a quantum of "frozen" light. We know that the photon is capable of moving inside the atom transiting between electron and proton so as to allow the atom to remain stable. We need to determine whether the photon can remain confined to be stationary in an SFT sense or alternatively whether it can move within a restricted range where the energy density allows it to remain stable inside a particle structure either inside the atom or perhaps inside smaller composite particles. Are there regions inside the electron or the nucleus for instance where its motion is confined in a stable and predictable fashion? We notice that any motion of the photon in free space is a dipolar motion and not a direct effect of EM. At first sight this appears contradictory since we know the photon mediates the EM forces as given by Maxwell's equations. However, any EM forces at the dipolar level of a composite photon may be a local or near-field EM effect between neighbouring photons that move in streams across the cosmos or between the ephectron, phroton, electron, and proton inside atoms as in SFT. So it is our object to determine if the photon can exist in stationary or dynamic confinement, for a start at least within the hydrogen atom.

Inside the atom there are two fields, the E- and the H-field. Each field is assumed to be maintaining the electron and proton in their principal mode. At this point in quantum theory the energy of the principal mode is assumed to be about $2\hbar\omega$, where $\hbar = 1.054589 \times 10^{-34}$ Js and $\omega = 1.033532 \times 10^{16}$ rad/s or −13.62 eV. In SFT each photon is assumed to transit between electron and proton via a series of resonant elastic collisions. The phase length during transit, $\pi/2$ maintains the periodicity of the atom providing a method for analytically comparing the energy of the photon with that of the electron

$$m_\gamma c^2 = \frac{h\omega_\gamma v_e}{4c} = \frac{q^2\omega_\gamma}{16\pi\varepsilon_0 c}$$

where ω_γ is the integer transit frequency of the photon within each cycle of the electron. For the atom to maintain its periodicity N_γ, the number of complete transits per atomic cycle, must be integer.

Using the value $N_\gamma = 54$, m_γ evaluates to 0.396×10^{-55} kg or 0.221×10^{-19} eV. This numerical value for m_γ is compatible with the experimental estimates for photon mass listed by the Particle Data Group. The analytic expression for the photon mass is compatible with the expression for the fine-structure constant[1]

$$\alpha = \frac{v_e}{c} = \frac{4m_\gamma c^2}{\hbar\omega_\gamma}.$$

Shown in Fig. 3.6 the atom can be better approximated by including the three quarks of the proton rather than the simple models where the proton is assumed to be a single particle.

Looking at the model of the photon in Fig. 3.7 each sub-photonic particle moves as a wave in the vertical θ-ϕ plane. Unlike the electron

---

[1]In SFT each photon is assumed to transit between electron and proton via a series of resonant elastic collisions. The phase length during transit, $\pi/2$ maintains the atomic periodicity providing a method for analytically comparing the energy of the photon with that of the electron $m_\gamma c^2 = \frac{\hbar\omega_\gamma v_e}{4c} = \frac{q^2\omega_\gamma}{16\pi\varepsilon_0 c}$, where ω_γ is the integer transit frequency of the photon within each cycle of the electron. In order for the atom to maintain its periodicity N_γ the number of complete transits per atomic cycle must be an integer. Ignoring any non-linearities, the electron's motion can be modeled as piece-wise linear. This collision-based form of periodic motion by the electron illustrates the non-classical, quantum nature of SFT. Assuming a polygonal motion circumscribes a circle representing the Bohr mageton, the photon collision frequency was estimated as **53** to several places of significance from the known value of the Landé g-factor assumed precisely known from both measurement and QED theory. More refined calculation may be needed to validate $N_\gamma$as the orbit is non-linear. The experimentally known fine-structure constant can also be used to estimate the solution as $N_\gamma = 54$. This was the solution chosen to estimate the photon mass.

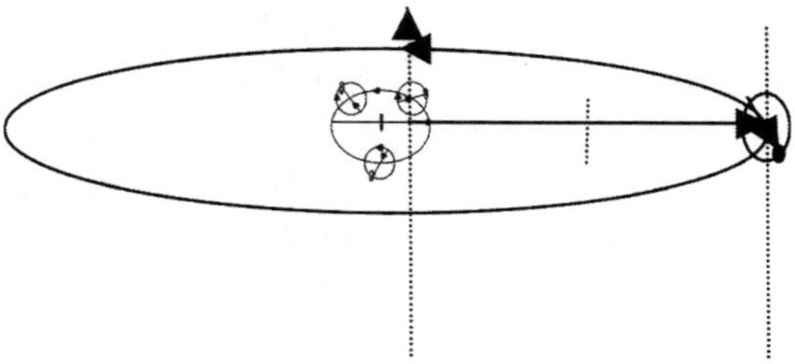

Figure 3.6. Composite model of the atom showing proton structure consisting of three quarks.

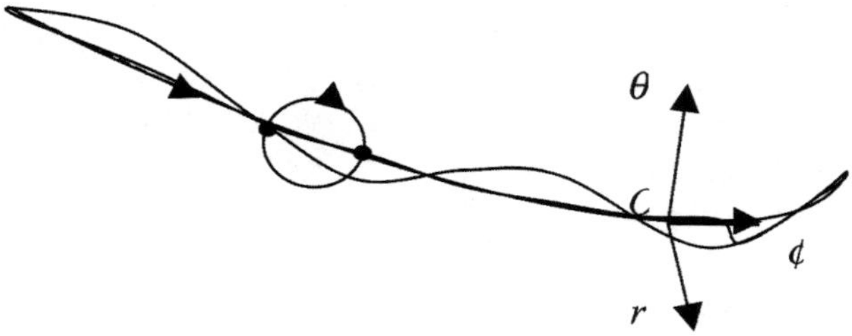

Figure 3.7. Hydrogenic photon moving in curvilinear free space past point O along ϕ axis rotating in the $\theta - \phi$ plane. Each sub-photonic particle moves as a wave.

or the proton that are charged monopoles, the photon is modelled as a dipole. In the limit of special relativity, the dynamics of the model becomes Cartesian as in Fig. 3.8. We can express the total energy of the composite photon as the sum of the two sub-photonic particles

$$E_{\mathrm{T}} = E_1 + E_2 \tag{3.5}$$

The velocities of the two particles are the sum of external and internal velocities, the speed of light c and the angular velocity v_i, where v_i is a bi-spinorial vector having components in the

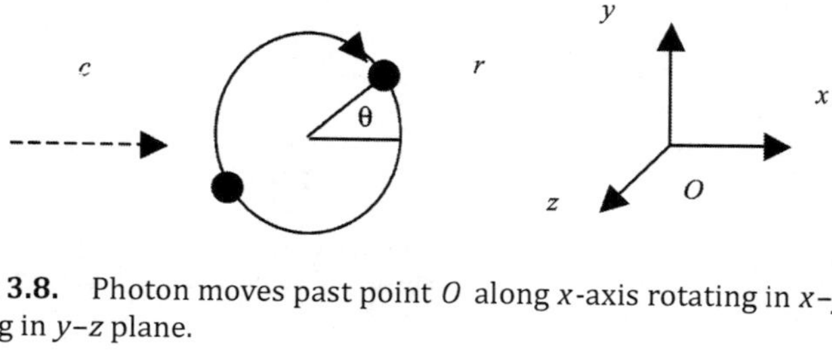

Figure 3.8. Photon moves past point O along x-axis rotating in x–y plane rotating in y–z plane.

orbital and cyclotron rotational directions. Given the masses of the ephectron and the phroton are both $m_\gamma/2$ they can both perform superluminal velocities as in (3.6) without violating relativity.

$$v_{\mathrm{e}} = c + v_i \cos\theta \tag{3.6}$$

$$v_{\mathrm{p}} = c - v_i \cos\theta \tag{3.7}$$

$$E_{\mathrm{T}} = \frac{m_\gamma}{2}\left[(c + v_i \cos\theta)^2 + (c - v_i \cos\theta)^2\right] \tag{3.8}$$

$$E_{\mathrm{T}} = m_\gamma c^2 \tag{3.9}$$

The total energy is equated to $h\omega_E$ where ω_E is related to the external wavelength of the photon, that of the EM field. For the case where the internal self-field energy of the photon is in an SFT balance with regards its internal energy where the spins of the ephectron and the phroton counteract each others' energy we may neglect the self-field energy and write

$$4N_\gamma \approx \frac{q^2}{4\pi\varepsilon_0 m_\gamma c^3}. \tag{3.10}$$

The factor 4 in $4N_\gamma$ relates to the phase length in this case 2π compared with the $\pi/2$ assumed inside the atom as the photon transits back and forth between the proton and the electron. This form is identical to the corresponding expression for the photon collision frequency as it transits inside the atom.

In the above, the ephectron and the phroton form a self-field binary system where the permittivity is that of free-space ε_0 and the charges are elementary units $1.602176487 \times 10^{-19}$ C. The two particles are locked together by EM forces while the photon in total is a dipolar particle being also affected by the dipolar forces imposed upon it at that particular point constraining it to move at relativistic speed c, in other words this dipolar force is similar to gravitation. The photon can also be treated as a stationary particle by assuming the ephectron and the phroton attract each other only via internal EM forces rather than any external dipolar forces; in this case, the energy density is not that of free space but much higher. Particle confinement was assumed to study the isolated hydrogen atom. If the external velocity is set to zero we may write an expression relating the photon energy and its internal velocities as in

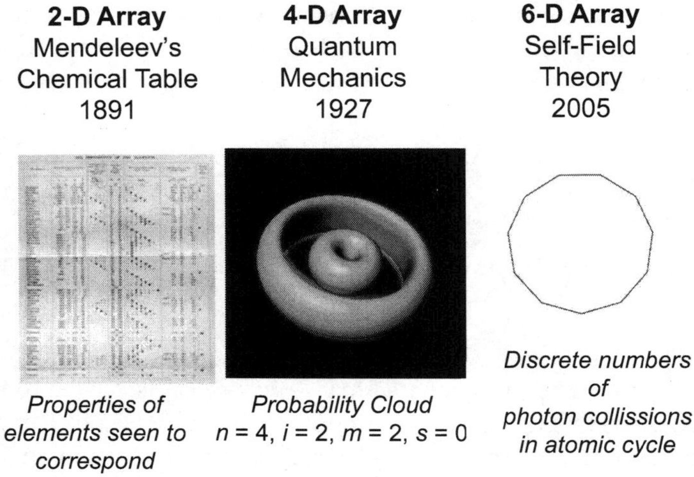

Figure 3.9. Historical perspective of the two new SFT quantum numbers associated with the resonance of the photon in its role as the binding energy in atoms and molecules. (Credit: Wikimedia.) See also Color Insert.

equations (3.3)–(3.4). Now the aim is to find what energy density will allow the photon to be confined as an SFT structure like the atom.

The historical perspective of quantum numbers changing with an emerging understanding of chemistry over the past 120 years is shown in Fig. 3.9.

3.3 Spin and Elementary Particles

QFT understands quantum spin as an intrinsic quantity if a particle is elementary. On the other hand SFT sees only a fractal physics, with no elementary particles in sight, on either end of the physical spectrum. Depending on the type and extent of the self-field effect SFT may see the particle spin as an intrinsic or a mutual effect. The internal cyclotron spin of the photon illustrated in Fig. 3.7 may be considered an intrinsic effect between the ephectron and the phroton. In the case of the electron (Fig. 3.6) it does perform a cyclotron motion analogous to quantum spin but this motion is a mutual effect between the electron and the proton. Since SFT treats the bosonic fields as fractal to at least the sub-photonic level, it appears all particles including the electron may also be fractal. As we have seen in Section 3.2 SFT indicates that leptons such as electrons may be composite and not elementary. Although the electron is well modelled as a point particle to around 10^{-22} m it too may have a composite and stable structure. If it does have an intrinsic spin it is due to its internal composite nature.

One of the greatest achievements of modern physics was the determination of the Landé g-factor for the electron spin magnetic moment where g was calculated using QED and measured to many significant places of accuracy to be $g = 2.002319304386$ in agreement with each other. Since this parameter is a function of the area of the electron's motion, its truncation error is quadratic in form and therefore much more accurate than parameters where the functionality may depend on distance of order 1. Nevertheless it is only in the light of SFT that we can understand that the numerical value of this factor is caused by the piecewise nature of the electron's actual motions. The orbital and cyclotron motions of the electron are a mutual effect and depend as much on the motion of the proton as on that of the electron itself. The nature of a particle's spin is due to the type of interaction whether mutual or intrinsic, monopole or dipolar. Bosons such as the photon may be "captured" by other particles to form energetically different particles and even different particle species. An example of this is the flipping of the magnetic north and south poles of the Sun in our Solar System. As the Sun's internal energy changes it passes through transition frequencies for the photon and this induces differing photon states with varying spin and radial states of the photon given by equations (3.3)–(3.4).

Pair production of an electron and positron occurs if the photon energy is raised above 1.022 MeV. This energy can be supplied by passing the photon near to an atomic nucleus causing the nucleus to recoil. At this energy the external frequency of the photon is around 2.4×10^{20} Hz and its external wavelength is 1.25×10^{-12}m. Within the photon the external energy may be approaching the internal energy given by (3.5) causing a reaction with the self-energy. This energy appears to be acting like the ionization energy for the photon where the ephectron and the phroton are a low-energy form of electron–positron pair inside the self-field dynamic balance but are energized into an electron–positron pair. At the same time the masses of the electron, positron, and photon indicates there is a relatively stable higher energy configuration for the isolated electron and the positron and a stable configuration for their combination inside the photon at lower energy levels. The energy of the field is transformed into a stable structure at higher energies in contradiction to our experience of particle physics to the present. In atomic weaponry increasing the energy density induces instability of the nucleus where long half-life particles follow.

3.4 Electron–Proton Shells and the Pauli's Exclusion Principle

Well-known to atomic physics is the shell structure of the outer electrons in multi-electron atoms. In the original Bohr theory, a construction rule for shells was a maximum of two electrons per orbital in order of increasing orbital energy. In 1936 Madelung came up with a better matching set of rules. Orbitals are filled in order of increasing quantum numbers $n + l$; where orbitals have the same value of $n + l$, they are filled in order of increasing n. This gives the following order for filling orbitals: 1s, 2s, 2p, 3s, 3p, 4s, 3d, 4p, 5s, 4d, 5p, 6s, 4f, 5d, 6p, 7s, 5f, 6d, and 7p. In nuclear physics a modified form of the atomic shell model is used to describe a set of nuclear shells in terms of energy levels.

What SFT reveals is another shell structure for the weak nuclear electrons and their complementary proton shell structures within the nucleus. As shown in Fig. 3.10 an internal SFT balance can be achieved between a weak electron and a proton within the central core or spherical region of the nucleus. In this region elevated energy densities are present via the electroweak constitutive parameters $\varepsilon_{WN}\mu_{WN}$. What this implies is that the strong nuclear effect occurs within an annular region where the quarks and gluons move.

In the early days of quantum mechanics, Pauling suggested that the chemical bond was due to a resonance phenomenon. In retrospect we can see that despite his insight, and his brilliance in using quantum mechanics to obtain a wealth of chemical information,

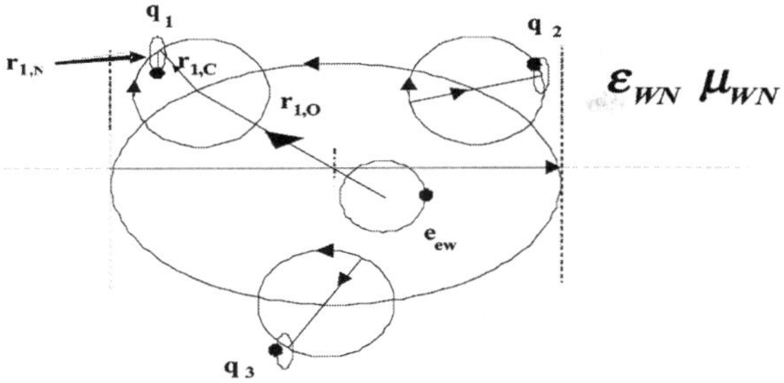

Figure 3.10. Electroweak and Nuclear shell structures. Analogous to the EM electron shifting shell as energy density is raised via $\varepsilon_0\mu_0$, so too the weak electron moves its inner shell if $\varepsilon_{WN}\mu_{WN}$ is raised, at the same time the proton shifts its nuclear shell.

the method itself was not capable of determining the exact bond relationships due to Heisenberg's uncertainty principle (HUP) and the lack of a complete EM field form. The history of both quantum mechanics and atomic chemistry are organic with numerical and other approximations used to aid the progress in atomic chemistry that was then blossoming. That there were errors in quantum mechanics was not considered, although Condon pointed out the limitations including singularities and numerical intractability that were becoming apparent. EPR pointed to the incomplete nature of quantum theory.

Atomic and molecular bonds are described by using orbitals, orbital diagrams, and electron configurations. As discussed the orbitals are filled in order of the lowest potential energy for the atom. The maximum number of electrons in each orbital is two according to the Pauli's exclusion principle. According to Hund's rule, orbitals of identical energy are considered half filled with parallel spin before they are paired. Such procedures are approximations used in part to avoid the need for quantum calculations where quantum mechanics is numerical and intractable for many-electron problems. These recipes and their quantum mechanics numerical counterparts have grown somewhat organically since the first days of chemistry as properties of various molecular substances have been studied via numerical calculations.

The Pauli's exclusion principle goes back to the early days of chemistry when it was realised that atoms and molecules with even numbers of electrons were generally much more stable than those with odd numbers. In 1919 Langmuir suggested that the electrons might be clustered in some fashion explaining features of the periodic table. Groups of electrons were thus thought to be gathered into shells. Bohr modified his theory to account for special numbers of electrons $(2, 8, 18, 32, \ldots 2n^2)$ that corresponded to "closed" shells. Pauli, following discussions with Bohr, realised that the number of electrons could be defined using four quantum numbers. For this purpose he proposed a new two-valued quantum number that was later identified as electron spin. For the electron $s = 1/2$ gives the total angular momentum of the electron as $S = \sqrt{\frac{1}{2}\left(\frac{1}{2}+1\right)}\hbar$ corresponding to the two possible spin states "spin up" or "spin down" $S_Z = \pm\hbar/2$ for the electron.

In SFT terms the same principle can be seen to be related to the way that electrons and protons orbit as EM binaries in a stable double rotation (see Appendix A). If a second pair consisting of another paired electron and proton are displaced in phase by π radians as shown in Fig. 3.9 the combined motion and structure is also stable. Underpinning this is a physical principle in the way the photons cannot stream between more than two charges particles at the sub-atomic level simultaneously in a stable fashion. While photons can transit between a particular electron and its paired proton as a pair of streams, they cannot move (orthogonally) between any other electrons or protons. This is a fundamental realization that greatly simplifies the determination of atomic and molecular structures. Thus the four EM particles shown in Fig. 3.11 are connected by only two (pairs of) photon streams, two E-field streams and two H-field streams that must rotate in synchronous fashion with the electrons and protons.

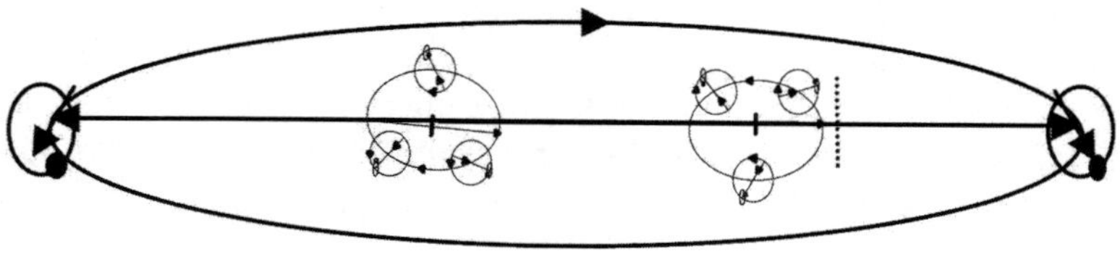

Figure 3.11. Pauli's exclusion principle: hydrogen molecule modelled as two protons with composite quark structure and two electrons.

3.5 Boson Structure

In Section 3.2 the hydrogenic nature of the composite photon was discussed. In atomic physics various elemental forms of electrons and nucleons of overall varying atomic masses form into the elements. These elements from hydrogen to hassium and beyond can be related to each other via their energy band structure. Thus spectroscopy can be used to determine the chemical composition of matter. Now that we have SFT we may use the ML equations to search for the analytic forms corresponding to these structures in the same way that QFT has been used to estimate chemical bond lengths and electron densities. In a similar fashion, by finding energy levels or densities at which the different bosons form, bosonic structures may be found to be related in the same way to atomic structures. This is termed *photon chemistry*. We may use SFT as it applies to the photon to search for these bosonic structures. In part this is a numerical search scanning across the energy density by varying the constitutive parameters, e.g. $\varepsilon_{WN}\mu_{WN}$ (WN - weak nuclear region). Note that the energy is inserted via two parameters and not one. The theme of SFT as a mutual effect between two or more particles is revealed.

In Section 2.13 a modified system of ML equations was given for the strong nuclear fields and particles, the gluons and quarks. The absence of any extrinsic gauge symmetry that comes with the ML equations allows recognition of the way a 2D geometry with two fields, two particles, and two rotations all played a central role in the mathematics of the EM region. This suggested a modification to 3D involving three fields, three particles all performing three rotations for the strong nuclear region. It appears the nuclear field mentioned in Section 2.13 is in fact the acoustic, vibrational, or longitudinal field. By looking at the way an electric and a magnetic field are transverse to motion the acoustic field will be in the direction of motion. By combination of a photon and a phonon we can see the underlying elements of a gluon, the boson for mediating the nuclear forces. A stoichiometric reaction can be proposed where sufficient energy density is present as in the nuclear region

$$3\gamma + 3\varphi \leftrightharpoons 4g \tag{3.11}$$

where γ is a photon, φ is a phonon, and g is a gluon.

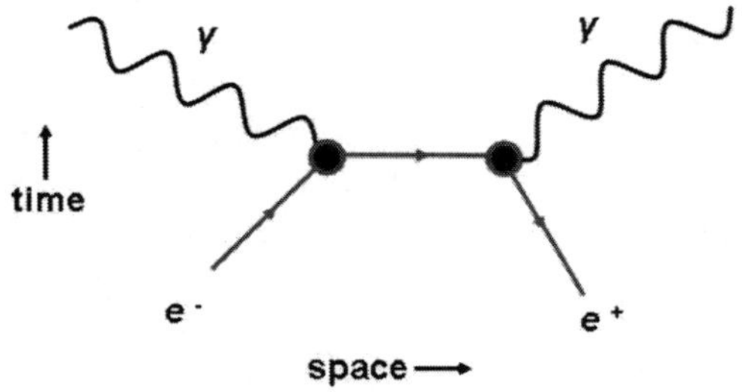

Figure 3.12. Anihilation of electron positron pair. (Credit: Wikimedia.) See also Color Insert.

The composite structure of the photon given in Fig. 3.5 shows the sub-photonic particles the ephectron and the phroton. Annihilation of the electron and positron produces two gamma rays with energies 0.511 MeV (see Fig. 3.12). The reaction causes a restructuring of the sub-photonic particles, the electron perhaps consisting of two ephectrons, and the positron consisting of two phrotons. These four sub-photonic elements rearrange themselves into the two gamma rays. The reaction indicates charge on the ephectron and the phroton is $q_\gamma = q/2$, half the elementary charge.

Looking at possible structures for the electroweak bosons, particle physics tells us that the two W^+, W^- bosons occur at energies above 80.4 GeV/c^2 while the Z^0 boson occurs above 91.2 GeV/c^2. A process that may be present in the physics is a series of windows in which SFT resonance can occur. Such frequency windows are known from ion cyclotron resonance where ions move in a magnetic field. The W^+ and W^- bosons may have a substructure similar to the positron and electron (two protons and two ephectrons) except existing at higher energies. The Z^0 boson may be a neutral combination of four sub-photonic particles, two ephectrons and two phrotons. Like the hydrogen molecule illustrated in Fig. 3.5 that is formed by symmetric location of two hydrogen atoms, the ephectron and the phroton are locked in an EM binary, while another ephectron-phroton pair is symmetrically located to form the Z^0 boson. Thus, Pauli's exclusion principle holds for photonic compounds as well as atoms. We can suspect that there exist shell structures for bosonic energy states. Shells are in series following the Big Band nucleosynthesis.

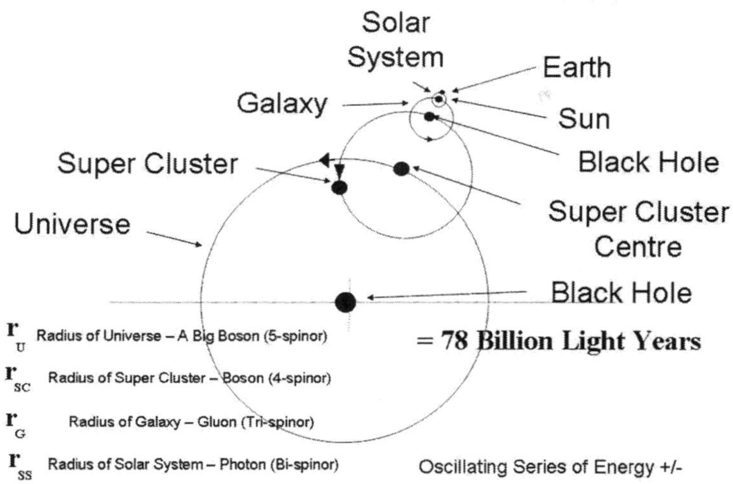

Figure 3.13. SFT model of the gravitational structure of the universe.

3.6 Gravitational Structure of the Universe

Following the discussion in Section 2.12 where a SFT model of solar system gravitation was tested the model of the Universe shown in Fig. 3.13 is based on the four levels of gravitational structure observed to hold across the visible universe. It is also based on the fact that there are distinct levels of gravitation given by SFT; each one a balance between matter and fields of different types. The four levels depicted are Universe, Super-cluster, Galaxy, and Solar System. Each field in the layered or fractal sequence is a type of integral of a larger field in which it is imbedded while each mass moving in the smaller domain is a differential form compared with the mass acting across the larger domain. These differentials and integrals are linked to the analytic forms of the spinors used in SFT for the fields and masses. This causes the formation of stronger/smaller and weaker/larger forces acting across the Universe. In addition across the model, alternating positive and negative collations of energy (black and white holes) move near the centres at each domain. All masses receive and emit field quanta. There appears to be a link between the number of rotations and type of gravitation. This includes rotations greater than three. The model presents a kernel mathematics for the unification of physics, a *standard model* of SFT that combines cosmology and particle physics.

$$
\begin{aligned}
W_U &= S_U + V_U + K_U + B_U \\
S_U &= W_{SC} = S_{SC} + V_{SC} + K_{SC} + B_{SC} \\
S_{SC} &= W_G = S_G + V_G + K_G + B_G \\
S_G &= W_{SS} = S_{SS} + V_{SS} + K_{SS} + B_{SS} \\
S_{SS} &= V_{SS} + K_{SS} + B_{SS} \\
V_{SS} &= -G\left(\frac{m_S^{SS} m_{Me}^{SS}}{r_{Me}} + \frac{m_S^{SS} m_V^{SS}}{r_V} + \frac{m_S^{SS} m_E^{SS}}{r_E} + \ldots\right) \\
K_{SS} &= \frac{1}{2}\left(m_{Me}^{SS} v_{Me}^2 + m_V^{SS} v_V^2 + m_E^{SS} v_E^2 + \ldots\right) \\
B_{SS} &= f_{SS}\left(m_\gamma^{SS} c^2\right)
\end{aligned}
\tag{3.12}
$$

The model sees energy as either extrinsic or intrinsic within each domain. Two forms of extrinsic energy exist, potential energy V and kinetic energy K; W is the total energy, B the binding energy and S the self-field energy of the particles. In the following system the model is simplified having only one sub-particle at the next level. The kinetic energies at each level are assumed to maintain a cosmological balance. As all motions are evolving the balance is only approximate and there should be a parameter ($\alpha < 1$) for each mass particle, e.g. planets as the orbital and cyclotron energies differ in each case depending on the actual evolution. The orbits are illustrated as simple circles, but they can be irregular and chaotic depending on the actual dynamics, on the fields and their ability to suddenly transit states. This includes the case where magnetic poles *flip* according to the photonic energies involved. Since photonic energies are continuous there may be flips as the photonic transition energies are crossed and the photon state changes polarity.

Table 3.1. Bosons and Masses at Each Level

Gravitational Level	Boson	Mass
Solar System	photon	Outer shell electron and paired proton (EM)
Galaxy	Gluon	Quark (SN)
Super Cluster	Gluon?	4-Spinor
Universe	Gluon?	5-Spinor

The bosons and masses at each level appear from the mathematical and observational evidence to be as shown in Table 3.1.

3.7 Photon Transit Across the Cosmos and Cosmic SFT Binaries

Analogous to the motion of the photon inside atoms, photons also transit between matter within the Sun and the Earth in dipolar or differential forms of SFT as long-range dipole–dipole forces called gravitation. To be discussed with regard to galaxies in Chapter 4 various other bosons form a distinctive fingerprint of gravitational structure within our Universe. Bosons of different kinds, ordinary EM photons, and strong-nuclear gluons appear to be involved at different levels of mediation within the Universe. The photon appears to have a special elementary role within the Universe capable of existing either as a particle or acting as streams to form fields. Thus, it can either act to be radiated across the Universe or being integrated into structures that are involved as elements in dynamic balances at various levels across the Universe. If multiverses exist photons may transit between universes. Alternatively Bose–Einstein condensates, "frozen light," may form at remote inter-universal locations.

As it radiates throughout the Cosmos, a photon may lose energy as it leaves the neighbourhood of a mass such as the Sun and the energy density decreases. Then, on approaching its destination, another mass such as the Earth, it regains energy as the energy density begins to increase near the Earth. This is demonstrated as a layered structure within the Earth's atmosphere. The energy within each layer is distinctive with a discrete or quantum state associated with the photonic state within each layer (see Fig. 3.14). These energy bands are continuous since the photon has a continuous eigenstructure. Distinctive chemical reactions occur within each layer such as cloud formation, and ambient temperature varies across height.

As an elemental form of matter an individual photon can move in keeping with SFT without losing energy along an enormous circular path of near infinite radius (or near zero curvature) at constant

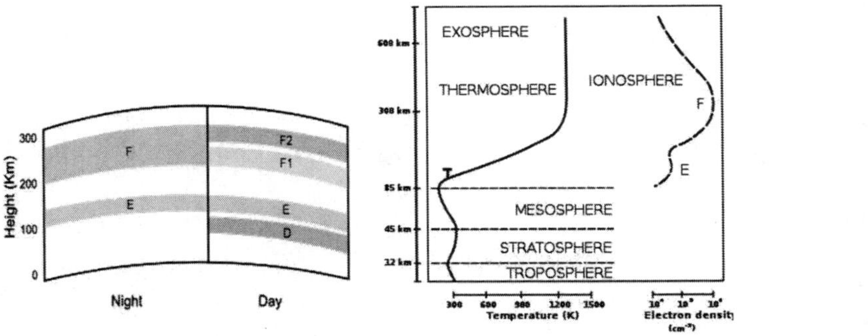

Figure 3.14. Layered structure of atmosphere and ionosphere. See also Color Insert.

velocity in accordance with Einstein's postulate. The photon is seen as one element in an SFT binary where the mass of the cosmos is the other element of the binary. Instead of an EM monopole binary, they form an EM dipolar binary outlined in Table A.1. The motion of any photon when moving in this fashion can be described as lossless even if it has non-zero mass (see Fig. 3.15).

A photon will move along a curvilinear path seeking out a loss-free path in relation to the centre of mass of the cosmos. While the photon does not collide with any other matter within free space it will move at its maximum speed which is related to the photon's mass and energy. As it moves into regions where the energy density is higher than deep space its overall speed is retarded. In the same

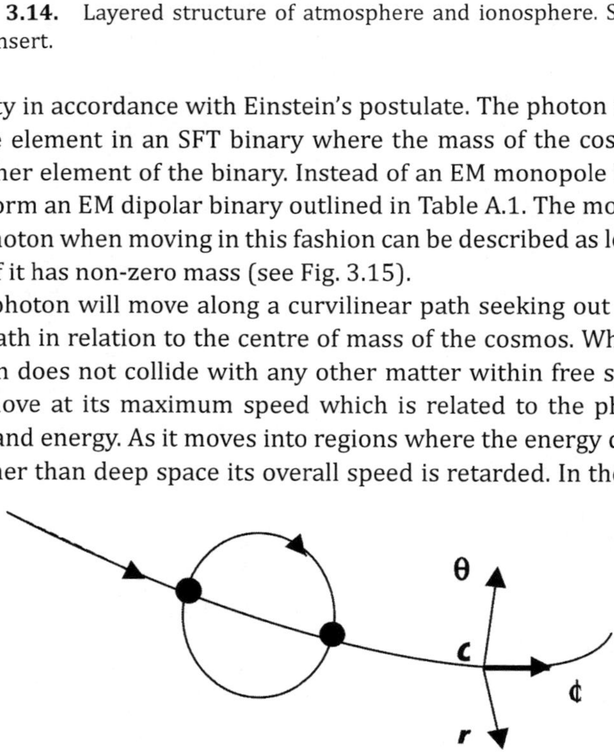

Figure 3.15. Composite photon moving in free space past point O along ϕ-axis rotating in the θ–ϕ plane of giant cosmic circle at constant speed without losing energy in accordance with GR. The photon also rotates internally in both the θ–ϕ and ϕ–r or r–θ planes in accordance with SFT. Two levels of SFT are involved in its overall motion.

way a phonon whose cross-sectional area is greater than the photon will move with a slower speed in the atmosphere associated with the terrestrial domain. Presuming it has the same mass as the photon, the phonon can also move with constant relativistic speed around a giant circle within the cosmos as it reaches deep space. The difference between the photon and the phonon is their polarity, their respective cross sections and hence their collision rates. In the case of the photon it moves sideways or crab-like through the cosmos, while the phonon moves facing straight ahead. Hence the phonon's frequency of collision is much larger than the photon.

Chapter 4

The Phonon

Production and appreciation of sounds via speech and music have long been intimate human abilities. Since ancient times acoustics has been a rich source of scientific research. Effects due to the speed of sound are more detectable using classical experimental methods than the much quicker speed of light. The invention of the phonograph and the prediction of acoustic surface waves are testament to the role played by acoustics within physics since the mid-1800s when Maxwell's equations thrust light and eventually the photon into the research spotlight. In comparison acoustics until recently remained a classical wave theory. The wave-particle duality of light resulted in quantum mechanics and the quantum field theory, but the wave theory of sound dominated in the absence of any corresponding particle theory of sonar phenomena, the phonon being a recent development within physics. During the 20th century various interactions have been discovered to exist between electric, magnetic, and acoustic sources. At this time self-field theory predicts both a theoretical and a physical unification of the phonon and the photon. In self-field theory the phonon and the photon both have internal structure and can react to form the gluonic fields that mediate between the up and down quarks inside atomic nuclei. Both the photon and phonon participate in a modified system of Maxwell–Lorentz equations involving the three fields, E-, H-, and acoustic

Self-Field Theory: A New Mathematical Description of Physics
By Tony Fleming

www.panstanford.com

(A-) fields. Piezoelectricity, sonoluminescence, the production of lightning, wind, and thunder by storm clouds, and the dynamics of suns within galaxies all support this three-field theory. Biodiversity is observed to be cyclic within the motion of the Milky Way and may be a resonance phenomenon associated with the galactic energy density.

4.1 Introduction

A phonon is a quantum of vibrational or acoustic energy that occurs either in solid state crystal lattices or in other media including air and water. Phonons can modify the electrical, magnetic, elastic, and thermal properties of gases, liquids, and solids. They play an important role in surface phenomena such as Rayleigh waves that travel within the surface of the Earth during earthquakes. In loudspeakers photons are converted via electromagnetism (EM) into vibrational or mechanical energy as phonons that are radiated acoustically. Brillouin scattering is an inelastic scattering of photons by acoustic or vibrational phonons that creates emitted phonons and alters the energy of the scattered photons; the phonon energy, imparted to the photons, can be measured by a spectrometer. This is the basis of confocal Brillouin microscopy that can measure the elasticity of biological tissues. Optical fibres can be similarly stimulated to provide laser gain. Phonons have a role in man-made energy cycles such as the combustion engine, and the rocket (see Figure 4.1), and natural energy cycles such as the weather and the cell cycles.

As optics relates to the visible photon spectrum, acoustics is a term that similarly applies to the audible phonon spectrum.

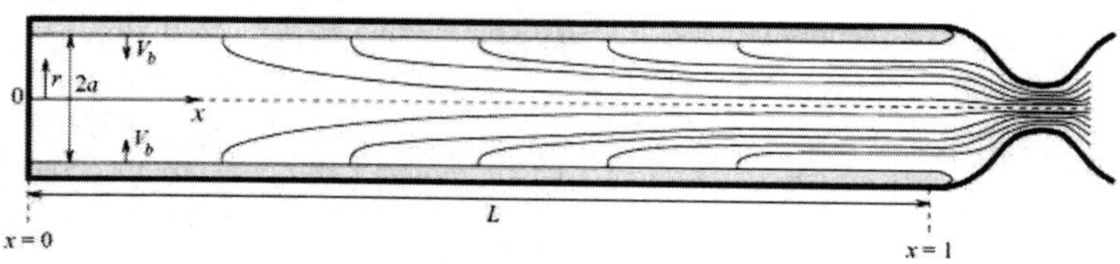

Figure 4.1. Rocket analysis of longitudinal acoustic velocities. (Source: Garcia-Schafer, J. E., Linan, A., *J. Fluid Mech.* 437 (2001), 229–254.)

Further as (EM) covers radiation over the photon spectrum, sonar covers radiations over the phonon spectrum. Vibrational energy is a macroscopic contact effect similar to contact electric currents while acoustic energy is delivered like EM as a form of radiation to an object via a medium such as air or water. Elastic effects such as stresses and shears are coherent phonon effects while incoherent or wide-spectrum phonon effects are thermal in nature. EM anechoic regions such as blackbody cavities including the embryonic universe have their analogue in acoustics where anechoic regions baffle sonar energy within the walls to minimize reflections and transmissions. It is tempting to classify the entire spectrum of sonar effects as all physics bar nuclear and EM, yet the same can be said for EM and nuclear physics. Each level of existence has different types of forces coexist in a balance across the fractal universe.

Science is divided as to whether the development of language was a relatively slow evolution or a sudden "genetic mutation"; modern man is thought to have been able to speak for 50,000 years or more. While the voice is produced by soft tissues, bone fossils of early hominids, apes, and modern man have been compared. In general across species the thorax houses the nerves that control sound production (see Figure 4.2). The thorax is much larger relative to body weight in modern humans than in apes. Neanderthals and early *Homo sapiens* correspond to modern humans, while *Homo ergaster* and the Australopithecines do not.

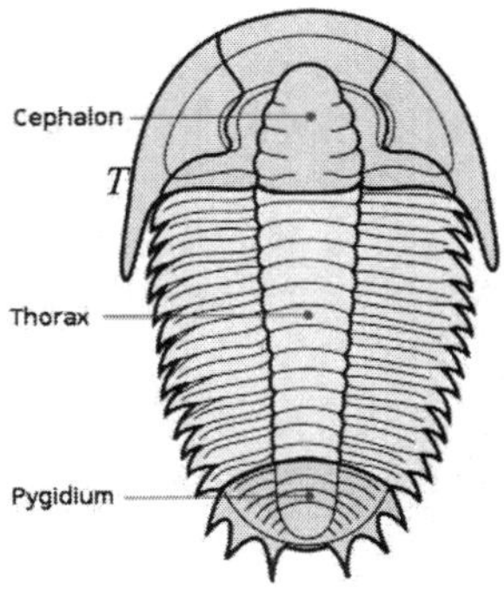

Figure 4.2. The trilobite body; cephalon, thorax, and pygidium. (Credit: Wikipedia.) See also Color Insert.

The study of speech as a sequence of standing waves within the human anatomy goes back to Pythagoras (569–475 BC). He discovered a mathematical relationship between pitch and harmonics. He believed in the "harmony of the spheres" by which planets and stars move according to harmonic notes. Although he is often accused of mysticism we now see that the idea of "cosmic music" is in part close to the way self-field theory (SFT) sees gravitation. Phonic forces may be involved in nuclear reactions at the cores of the planets and stars; astrobiologists are looking at the diversity of life across galactic motions. Around 430 BC, Hippocrates the "father of medicine" studied anatomy including the function of the larynx. Newton at the age of 24 performed a simple experiment noting that on filling a tall glass with liquid the vowels u, o, a, e, and i could be heard. This process mirrors the action of the glottis that acts as the source of voice production. The hypothesis that resonance was involved is due in 1681 to Henrici who suggested vibrations of the vocal cord were responsible for phonation. Dodard in 1700 discovered that voice pitch was due to tension in the vocal chords. Ferrein in 1741 studied larynxes of dog, pig and cow cadavres.

The modern study of speech production and recognition springs from this long history of the study of sound and speech. Some studies have conducted trials to see if speech can be correlated with underlying patterns. Computer simulations represent acoustic signals as trajectories in 2D space, a "hill climb" (see Figure 4.3). Each seemingly random step is kept if it represents an "improvement" or else rejected. Improvements take various forms, e.g. "distinctiveness." Experiments on phonetic languages support self-organization into patterns other than the classical Darwinian evolution strategy. Yet these trials fail to answer a critical issue as to any underlying physical nature of the improvement. Where acoustic forces are present, what is the size and position of the standing waves as a function over the history of the bones and soft tissues involved in voice production? Have the Earth's E-, H-, and A-fields been a factor in the changing size of the vocal organs over cosmological time scales?

The role in evolution of Earth's E-, H-, and A-fields, as a function of location on the Earth's surface and over cosmological time deserves research attention. In the Earth's early history the Moon was closer

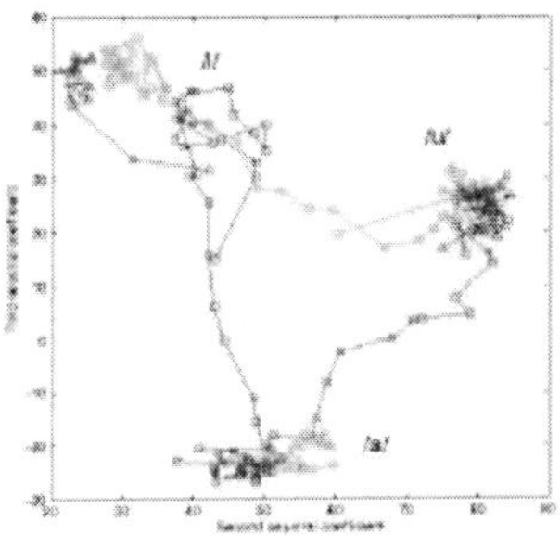

Figure 4.3. Trajectory representation derived from recorded acoustic data of Dutch speech. (Source: Zuidema, W., de Boer, B., The Evolution of Combinatorial Phonology, University of Amsterdam Institute for Logic, Language and Computation, 2006.) See also Color Insert.

to Earth than today following the "Big Splat," an event hypothesized to be a collision between the Earth and another planet, the origins of our Moon. Similarly laser measurements show the Earth is moving slowly away from the Sun. The geo-electric and -magnetic forces in prehistoric times would have been stronger than they are today. The primordial record shows various life forms to have been generally bigger than today. Fossil trees, plants, insects, and mammals were often larger than their modern counterparts. Oxygen levels in the atmosphere were one important factor in determining the size of a species. Ambient acoustic and EM energy levels on the surface of the Earth were also larger. Only whales that were somewhat protected during the atmospheric extinction event 65 million years ago are bigger than their associated prehistoric forms. In fact modern whales are the biggest life form ever known. Perhaps E-, H-, and A-fields are an underrated element in the genetic mix. DNA is known to be sensitive in ambient energy levels.

Man's ability to discriminate and appreciate sounds including language and music appears in the artefacts of pre-history as do the other human psychological functions including the innate religious, artistic, and mathematical understandings. Sound it seems is a deeply integrated element of what it has been and continues to be human. It is likely that music is stored neurologically as a long-term form of memory rather than as a short-term memory. Sufferers of Alzheimer's disease and dementia are helped to stave off the effects of memory loss by music therapy. Even people deep into the disease near death can be helped in this fashion. Present-day musicians use their creative abilities to write peaceful, soothing music as a form of therapy for various neurological illnesses in a similar way to the use of massage to relax the physical body. It is doubtless that the popular music of the 1960s brought about great changes in society including a freeing up of social restrictions. This was a political factor in the U.S. during the Vietnam War and in the former USSR during the 1970–1980s before the Berlin Wall came down. A similar process is occurring within the East and the Middle East today. Indeed as with art and science, the process of creating a shared music by a number of players and listeners is a form of communication between people of different cultures, creeds, and languages permeating societies across the world. This modern form of social communion can be traced to the music of tribal Africans who were traded across the Atlantic as slaves almost since the discovery of the New World. The modern trend towards globalization has its roots in this shared musical heritage.

The use of sounds including music as a means of changing the psychological state of the individual or society is not a recent phenomenon. Ancient peoples invented and used musical instruments not only for long-range communications, but personal entertainment and religious ceremonies. Conch shells, antelope horns, bamboo and bone flutes, cymbals, drums, lyres, lutes, bull roarers all dating from pre-history demonstrate man's early attachment to musical sounds. Singing bowls that generate a mainly monotone resonance are known from around 800–900 BC and in Buddhist practice are still considered a transforming agent used to aid meditation, prayer and chanting. Medieval music including chants illustrates the connection between music and religious ceremony that is found throughout history. Musical notation has a history and prehistory comparable to that of written language and mathematical notation. Sound has also been used from ancient times as a modality for delivering medical therapy. The traditional Chinese medicine Qigong in its modern electro-acoustic form is applied as a vibrator when directly contacting the body and as an acoustic device when no surface is directly stimulated. Qigong is said to be

Figure 4.4. Jadeware with description of Qigong from Warring States Period 476–221 BC. (Source: http://tcm.chinese.cn/en/article/2009-08/24/content_11271.htm) See also Color Insert.

the forerunner of Chinese acupuncture that dates back to the time of Confucius (see Figure 4.4). Acupuncture can be used at the points of low resistance on the body surface, the acupuncture points. Qi, or Chi, is the form of subtle energy that is thought to permeate the universe including the body. Eastern medicine seeks to balance the opposites Yin and Yang within the body. These concepts correspond to the modern scientific concepts of the photon and electric current. In the corresponding traditional Indian medicine, chakras are centres of energy within and around the body, like the modern concept of body-field whereby from bioelectromagnetic studies there is now known to exist a biogenic field that extends beyond the physiological body parts. Unfortunately like the ancient Western science of astrology, these early forms of scientific knowledge were embroiled with superstitious and fatalistic concepts. Feng Shui and the selection of camp sites by North American Indians, reveals an early understanding of electric, magnetic, and acoustic subtle energies that modern science itself is only now coming to understand. At the same time it appears that it is the scientific method that allows progress past the barrier of superstition that attenuates knowledge. Acoustic and vibrational therapies and frequency-based more general forms of Qigong are being successfully used with violent young offenders to bring about psychological therapy. Vibrational and acoustic energies of varying frequency are being used as therapies for various physical ailments

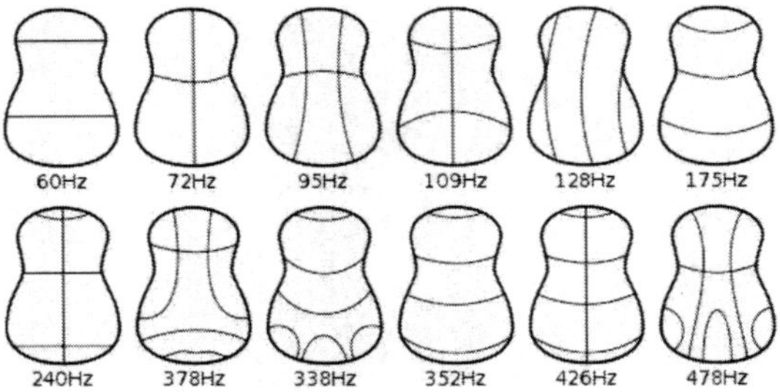

Figure 4.5. Visualization of guitar modes by Chaldni. (Credit: Wikipedia.)

including surface blemishes and deep tissue ailments. The modern form of acoustic therapy is applied as a resonance to tissue lengths to destroy pathogens or to promote wellness, e.g. blood perfusion within the body.

In terms of the resonance modes or eigenvalues of acoustic phenomena there is a rich vein of experimental effort reaching back to Hooke the contemporary of Newton in the 17th century. In addition to inventing the optical microscope, Hooke performed various acoustic experiments including the invention of an ear trumpet in 1668 to assist the deaf and hearing impaired. He demonstrated this device to the Royal Society. To transmit music and speech over distances he demonstrated an early form of telephone consisting of two tin cans connected by a piece of wire; Hooke is thus considered the founder of telecommunications. To demonstrate acoustic modes, Hooke used a flat circular glass dish covered in flour to visualize the vibrational modes within it as a violin bow was pulled along its edge. Chladni (see Figure 4.5) a century later performed the same experiment using a metal plate covered with sand.

Looking at the basic physics of early flutes and stringed instruments, the length of the chamber or string is related to the pitch; there is also a relationship between higher octaves and half-lengths. It was Pythagoras who developed the diatonic scale, the foundation of modern European music. He saw a relationship between music and the integer numbers, an early recognition of the quantum nature of physics. Unfortunately a compendium of esoteric knowledge including numerology, astrology, and alchemy also sprang up East to West during this era. The four ancient elements, earth, air, water, fire, their qualities, the seasons, phases of the Moon, their links to growing and harvesting crops added to an early "folk science" that emerged from an intuitive but faulty reasoning. The classical scientific era finally ended in Alexandria when the library burnt down in the 7th century AD, the last of a sequence of fires. Not until experiment and theory combined several centuries later would science again progress.

Early nomads such as the Australian and African hunter-gatherer aborigines, and the North American Indians, had probably until recent times lived the same way of life for many millennia perhaps even during and before the last Ice Age. It can be said that while modern man and his 20th century technology may have brought the world to the edge of ecological disaster, his forebears obtained their needs from their surroundings in a completely sustainable way. From studying the way these elder races use their natural environment deciding where to live and hunt, they appear to have an intuitive understanding of E-, H-, and A-fields (see Figure 4.6) and how these fields affect them on a day-to-day level. It may be that city life that evolved from earlier times in order to give protection to the inhabitants and that in the modern era has provided infrastructure

Figure 4.6. Didgeredoo. (Source: http://medgadget.com/archives/2007/01/do_you_have_sle.html) See also Color Insert.

and social support may in time cease to be so heavily populated relative to the rural surrounds. This will be the case if energy can be provided outside of the modern energy grid and sufficient internet-based social support becomes available. Urban problems including carbon footprint, noise, chemical, electromagnetic, and atmospheric pollutions need to be addressed. If as appears the case, sonar energy is a key to unlock new methods of nuclear generation the future may indeed be a more sustainable mix of technology and social support away from urban centres.

4.2 Brief Outline of Recent Acoustic and Vibrational Research

The invention of the phonograph by Edison and the phone by Bell is testament to the major role of acoustics since the mid-1800s. During the American Civil War, the use of telegraphy in which the signal was encoded via Morse code became widespread. By converting the sound into a form suitable for recording or transmitting both the phonograph and the phone were enabled. In the case of the phonograph a stylus was used to scratch a groove onto a cylinder that rotated mechanically. The cylinder was covered with wax or tin-foil and depending on the depth of the groove the pitch and intensity of the sound could be recorded. In the case of the telephone, the tone changes the electrical current in a wire.

In the late 18th century Galvani was dissecting a frog during an electrical storm. When he touched the frog with his metallic scalpel the frog's leg twitched as if alive. This demonstrated the bioelectric basis of biological life. Exposures of biological bodies to acoustic, vibrational, and EM frequencies have recently been recognized as an important means of medical therapy. Like holding a small twig under a microscope on a hot day, photons and phonons can be enhanced to cause biological effects to those entities in focus. The wavelength in tissue can be tuned to biological lengths such as the DNA of various tissues within the body or the icosahedral capsid coats of many viruses. These frequencies can be used either to promote or hinder biological processes including stages of the cell cycle, and various forms of cancer. Brain tumours for instance often spread more rapidly than normal healthy tissues and this replication rate can be utilized to preferentially attack malignant cells during replication. Therapeutic use of vibrational frequencies can cleanse the arteries of plaque buildup or destroy kidney stones. Efforts to use such monochromatic energy as a resonant mode of therapy for a range of treatments go back to the mid 20th century. Potentially more powerful than pharmaceuticals or surgery this focusing is a structural effect where use of resonance or specific photon and phonon states may be used to promote growth of healthy cells or destruction of pathogens. Civil engineers are aware of destructive resonance effects where bridges and other structures interact with wind shear forces causing destruction as the structure vibrates at its modal frequency. Other hybrid and non-resonant methods of acoustic surgery are being developed for instance at the Center for Industrial and Medical Ultrasound, University of Washington, Seattle. Focused high-power ultrasound can deliver high energy levels to specific sites deep within the body. The intensity at the focus is of the order 1000–10,000 W/cm^2 and can be focused to a region the size and shape of a grain of rice. National and international acoustic standards similar to those for public and occupational exposures to EM are embryonic and need development. Noise levels near highways carrying high traffic volumes and near wind turbines require urgent attention. Another issue is the effect upon marine life of naval and commercial use of marine sonar. Dolphins and other marine mammals may suffer bio-effects to ears and internal organs.

Surface acoustic waves were predicted by Lord Rayleigh in 1885 Rayleigh waves travel within the surfaces of elastic solids. The particles move around elliptical orbits in a plane normal to the surface along the direction of propagation. The particles move in the negative direction at around larger ellipses at shallow depths and in the positive direction in smaller ellipses below a depth roughly equal to the acoustic wavelength. Because the wave propagates in 2-D as distinct from waves propagating in the bulk medium it loses energy as a function of $1/\sqrt{r}$ where r is the radial distance. Looking at Fig. 4.7 the wavelengths in air and in liquid of atomic particles differ. Beginning at the media interface there is a series of quantum states where atoms and molecules assume a balance that

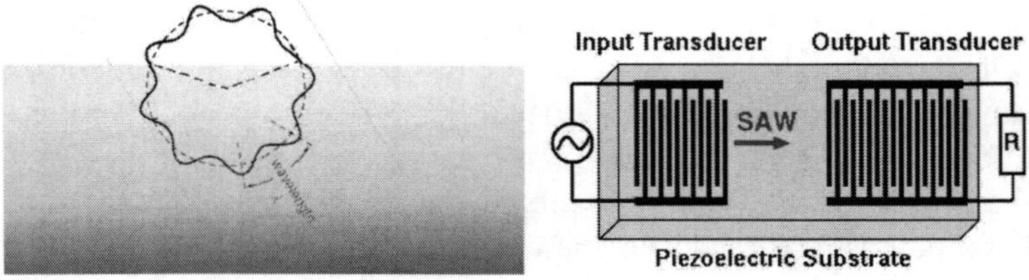

Figure 4.7. (a) Surface acoustics wave (b) SAW electronic component. (Credit: Wikipedia.) See also Color Insert.

changes with depth until the wavelength is essentially that of the bulk liquid. Such surface waves have become the basis of important applications in non-destructive testing, geophysics, seismology, and the production of electronic components.

In the theory of the centre-fed half-wavelength dipole antenna there are both transverse and longitudinal components of the emitted photons. In the mathematics of the dipole the EM components are filtered out from the complete radiated energy to obtain the far-field E- and H-fields. The remnant longitudinal components are similar to phonons. These are not nuclear phonons escaping from the nuclei of atoms within the wire antenna but photons that are not transverse to the direction of wave propagation. This type of longitudinal radiation is an electro-acoustic effect, a mix between transverse photons and longitudinal "photons," in other words phonons. In the dipole such phonons are sensed as heat, a form of inefficiency. In the reverse case phonons incident to a metallic surface can cause currents to flow in the surface. From the theory of the photon we see that a corresponding theory exists for the phonon the quantum of acoustic energy. As with the composite photon, a composite phonon can be illustrated as in Fig. 4.8.

Piezoelectricity is a contact phenomenon where solids such as crystals, ceramics, and bone, generate electric fields in the bulk of a crystal in response to directly applied stress. Phonons applied to a solid structure result in photons being transmitted within the crystal enough to cause an electric field or a potential to be detectable. The effect is elastic up to a deformation of about 0.1%. Such phenomena form an expanding piezoelectric industry including non-destructive

Figure 4.8. Composite phonon moving past point O along x-axis rotating in $y - z$ plane in accordance with SR.

testing. Piezomagnetic materials have also been discovered as observed in anti-ferromagnetic materials. The mathematics of piezoelectricity can be treated as a combined effect of the electrical and mechanical behaviour of a material

$$D = \varepsilon E \tag{4.1}$$

where D is the electric displacement, ε is the permittivity of the material, and E is the E-field strength, and Hooke's Law

$$S = sT \tag{4.2}$$

where S is strain, s is compliance, and T is stress. The effects couple as given in the following equation where vectors are indicated by curly brackets and tensors by square brackets

$$\begin{aligned} \{S\} &= [s^E]\{T\} + [d^t]\{E\} \\ \{D\} &= [d]\{E\} + [\varepsilon^T]\{T\} \end{aligned} \tag{4.3}$$

Serendipity has played and continues to play a most important role in science. In 1934 two German scientists Frenzel and Schultes serendipitously stumbled onto a type of electro-acoustic coupling now known as sonoluminescence while looking for a way to speed up the photographic development process. They found that a photographic plate immersed in water became dark indicating exposure to light when the water was being acoustically driven. They observed a faint bluish-white glow or luminescence emanating from bubbles within the liquid. In the later 20th century, the bubbles were related to the sound pressure cycle. The bubble locations were at the antinodes of the acoustic standing wave. Two types of phenomena called multi-bubble sonoluminescence and single-bubble sonoluminescence exist. Called the "hydrogen atom" of sonoluminescence the single-bubble phenomenon was isolated experimentally in 1989 (see Figure 4.9). The equations governing

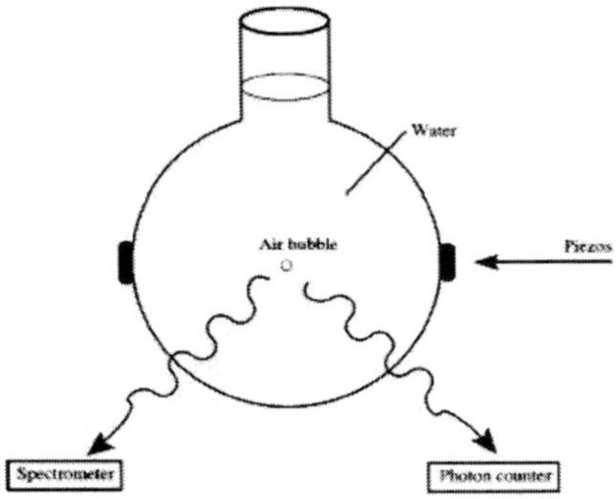

Figure 4.9. Sonoluminescence chamber. (Credit: Brenner, M.P., Hilgenfeldt, S., and Lohse, D., *Rev. Mod. Phys.* **74**(2), 2002).

the acoustic phenomenon of bubble creation in the liquid are the compressible Navier-Stokes equations:

$$\rho\left(\partial^t u + u \cdot \nabla u\right) = -\nabla p + \eta \nabla^2 u + \zeta \nabla \nabla \cdot u \tag{4.4}$$

$$\partial^t \rho + \nabla \cdot (\rho u) = 0 \tag{4.5}$$

where u is the fluid velocity, ρ is the density, p is the pressure as given by an equation of state, where η is the shear viscosity, ζ is the bulk viscosity of the liquid. Where σ is the surface tension within the bubble, the bubble radius $\boldsymbol{R}$ is a function of time, $\boldsymbol{R}(t)$, and can be written as

$$R\ddot{R} + \frac{3}{2}\dot{R}^2 = \frac{1}{\rho}\left(p_g - P_0 - P\left(t\right) - 4\eta\frac{\dot{R}}{R} - \frac{2\sigma}{R}\right) \tag{4.6}$$

Thus the dynamics of the bubble is complex and delicate, a problem of applied computational fluid dynamics. Various regimes of the mathematics exist. The single-bubble sonoluminescence solution can be divided into several distinct regions including an expansion, a compression, and several afterbounces at a resonant frequency depending on a number of factors including the presence of a small ratio of a noble gas for example argon. Using photomultipliers, the flashes have been linked to the frequency of the acoustic source and occur near the end of the compression period. During bubble compression, temperatures in the bubble rise dramatically causing ionization of some of the noble gas. While the computational

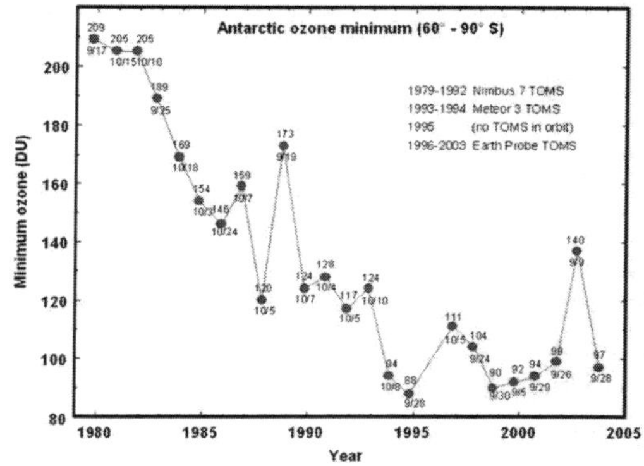

Figure 4.10. Ozone depletion, 1979–2004. (Source: http:/toms.gsfc.nasa.gov/multi/multi.html) See also Color Insert.

efforts track the approximate evolution of the bubble, they give no indication as to the source of the light flashes.

Understanding of the way storm clouds produce lightning, wind, and thunder has developed slowly over the past decades. An ever detailed picture is emerging of the complex relationship between Earth's weather patterns and the chemical reactions within the atmosphere including the recent observations of the ozone layer depletion (see Figure 4.10) via carbon deposits directly into the atmosphere. According to metrological definition there are seven layers surrounding the surface of the Earth: the troposphere, stratosphere, mesosphere, thermosphere, exosphere, ionosphere and magnetosphere. Storm clouds are found within the bottom layer, the troposphere wherein temperatures fall with increasing height. Up to around 15–30 km in height some clouds are tall enough to bump up against the stratosphere where ozone absorption of sunlight causes heat buildup and temperatures to increase with height.

Lightning is considered due to the breakdown of air as storm clouds gather. In this early stage, the clouds drop in height due to a drop in temperature as can occur after sundown or during nighttime. This causes the electrostatic buildup of charge between the Earth's ground surface and the underside of the cloud. When this causes E-fields of the order of MV m^{-1} the breakdown begins and a plasma of ionized air particles forms. This is a "sea" of electrons that are ready to find a route within the plasma to discharge the surface charge below. This surface charge is also ready to find a path to the storm clouds above. Hence two sets of seemingly random walks or leaders begin from above and below inching towards each other. Hence low-lying clouds and tall structures including trees and buildings are locations where these walks take place. High-lying regions around hills, mountains, and broadcast towers are also prone to lightning strike. The small steps are actually paths of least resistance, so areas of water and metallic structure can also be involved, and depend also on the dielectric nature of the ground. Eventually the leaders from above and below join up and the air explodes as electron charge flows from the clouds down to the ground. What now occurs is that electrical transients flow down the plasma and a shock-wave, or thunder, occurs. Thus energy runs down the plasma towards ground and upwards from ground towards clouds. Cloud–cloud lightning can also occur. Various forms of lightning exist including sheet lightning, red sprites, and blue jets that are plasmas formed between the storm cloud and higher regions within the ionosphere above the cloud.

The production of rainwater or precipitation is considered to depend on the "dew point," the temperature to which a parcel of atmosphere needs to be cooled before it becomes saturated. Below this temperature rainfall occurs when the condensed water molecules become disassociated from the condensation nuclei, airborne particles of dust, ice, and other molecules including chemicals such as carbon monoxide emitted by internal combustion engines. This condensation falls as snow, and melts to form rain as the temperature rises close to the ground. Hail and snow occur at ground temperature below the freezing point of water.

There is no scientific consensus about the origin of wind (see Figure 4.11). According to some, wind is thought to be due to differences in air pressure either in the horizontal or the vertical plane. Others define it as air in motion, due to differences in temperature. The ideal gas law is sometimes given in terms of the theory of wind formation. Apart from listing statistical records of wind speeds and directions, the theory of wind is limited at present (see Figure 4.12). What is known is that along with sun power, wind power is a

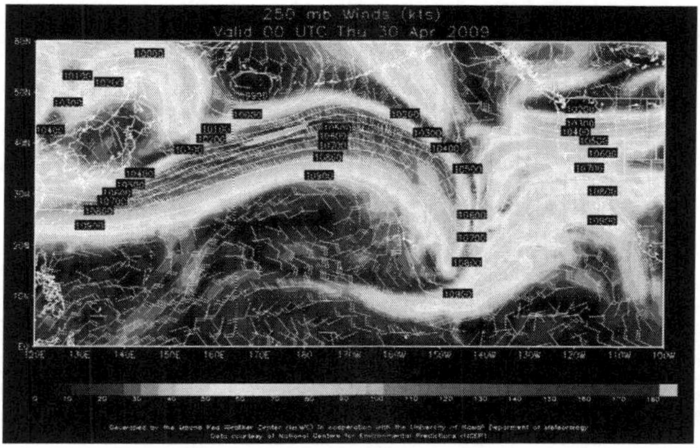

Figure 4.11. Wind forecast for North Pacific, April 30, 2009. (Source: Http://mkwc.ifa.hawali.edu/) See also Color Insert.

completely renewable supply of energy and the current focus of many scientific agencies. There are plans for off-shore wind farms. Events relating to climate change appear to be a strong factor towards a rapid growth of the embryonic renewable energy industry across the globe.

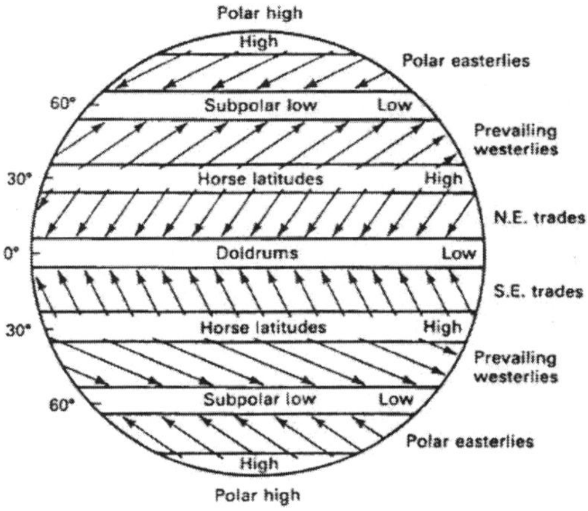

Figure 4.12. Theoretical global wind pattern. (Source: Chapter 2 Wind Energy System by Dr. Gray L. Johnson November 20, 2001, http://www.rpc.com.au/products/windturbines/wind-book/wind2.pdf)

Figure 4.13. According to SFT lightning, wind, and thunder are forms of electric, magnetic, and acoustic energy. (Source: www.stormscapedarwin.com) See also Color Insert.

One area of where SFT may elucidate possible nuclear interactions is the formation of lightning, wind, and thunder from cloud formations (see Figure 4.13). According to SFT, the storm clouds can be treated as large molecular aggregates. As the temperature and pressure vary with height, the atomic and nuclear binding energies within the aggregate may vary and changes in E-, H-, and A-fields occur. It is possible that variations in quantum states of both EM and nuclear binding energies drive the production of lightning, wind, and thunder within the cloud. The aggregates thus change their energy state giving off E- and H-fields as photons and A-fields as phonons. The energy associated with the E-field is emitted as a translational energy and drives the production of lightning. The H-field is emitted as a rotational energy that drives the production of wind. The longitudinal energy is released as acoustic energy that produces thunder. Time variations in both lightning and wind may also induce a transformation of E- and H-fields into A- fields and vice versa. Thus it may not only be the spatial variations of height, pressure and temperature, but also the time variations of energies associated with the cloud masses. These cloud masses are just large molecular aggregates that undergo energy state changes that are emitted or transformed into various forms of lightning, thunder, and wind. This is similar to the energy changes used in spectroscopy or the beats from a stethoscope.

Clouds may be monitored in a fashion similar to the measurement of energies in a spectrometer where photons of differing wavelengths are detected according to the binding energies of the material being burnt. A chemical disassociation occurs within the atoms of the substance being tested that releases photons of specific frequencies depending on the chemical states involved. With storm clouds the temperature drops instead of rising. The energies within clouds reach the point when a nuclear state change may occur in distinction to a chemical reaction. It may be possible to measure both photons and phonons, thus producing a "nuclear spectrometer" with which to measure the nuclear activity in the cloud. Such a device might be able to be used widely to determine the nuclear activity within atoms and molecules thus opening up the experimental door beyond the knowledge gained by particle physics where collision-based theory and practice provides partial knowledge of weak nuclear events via Geiger counters. Such nuclear spectrometers could be utilized to monitor seismic events such as volcanoes, earthquakes, tsunamis, tornados, and plate tectonics. They may be useful in determining safe working conditions at nuclear energy facilities.

One important piece of evidence for the tri-field hypothesis discussed in this chapter is that lightning and thunder appear to be produced simultaneously. There is a rule of thumb for measuring the distance a person is from a lightning strike by timing how long it takes to hear the subsequent clap of thunder. Each 2.9 s the light can travel approximately 1 km. In this calculation, the speed of light is taken as approximately infinite while at 20° C the speed of sound is around 343 m/s. If in fact the two forms of energy are produced at the same instant in time, then the strong nuclear equations given by (2.45) are supported as a hypothesis.

Returning to the experiments on sonoluminescence, these phenomena may be similar to the changes in nuclear energy of clouds at heights above the Earth. In the case of the single-bubble phenomenon atomic structures may alternatively take up and release energy within the strong nuclear regions thus emitting both E- and A-fields at the acoustic antinodes as a series of evanescing waves. A similar effect to sonoluminescence known as Cherenkov radiation is seen as a faint green glow due to the effect of radioactivity in liquids. These phenomena indicate a synergy between the photon and the phonon. Such production of spherical bubbles may be how cells first evolved within fluids on the surface of the Protoearth where strong E-, H-, and A-fields may have been present. At present the strong nuclear reactions are not recognized as being a part of our everyday life. Our experience of these forces has been either via atomic weaponry or high energy laboratories. Nuclear reactions appear in fact to be part of the fabric of the everyday physical world around us and in its biological life forms including our own bodies where short and long term memories may be stored neurologically via biophotonic and "biophonic" mechanisms.

4.3 The Phonon and Cosmology

The role of the phonon in shaping the physical universe goes back to its very first moments we now refer to as the Big Bang. Acoustic modes were present within the formless embryonic universe as it burst forth into time and space. Both acoustics and EM were the driving forces behind the gravitational structure within the universe including the birth and development of super clusters, galaxies and stars. Acoustic and EM modes within the early universe changed the otherwise random energy density across the universe. Recent advances in understanding the early structure and formation of the universe have come about with the accidental finding of cosmic microwave background (CMB) radiation in 1964 by astronomers Penzias and Wilson. They thought the isotropic hiss they had found was due to guano or pigeon droppings inside their horn antenna. Since then cosmologists have scanned this microwave radio energy for fluctuations initially using the Cosmic Background Explorer satellite (COBE) and more recently the Wilkinson Microwave Anisotropy Probe (WMAP). They have not only mapped the EM energy as it appeared when the universe was 380,000 years old but also its initial acoustic response, its initial "chord," as if it were a musical instrument. Thus a picture of creation is apparent of God and the universe as a musician playing a cosmic instrument.

In the same way that the initial space into which the universe was born was a perfect blackbody for EM or transverse waves, it was also a perfect anechoic chamber for acoustic or longitudinal waves. Only extreme energies were present, the size of the universe precluded long wavelengths. Within the early universe the conditions were such that EM and acoustic modes occurred and shapes and nodes within the ever-growing spherical space began to form. "Pure energy" begins to condense into nodes or particles of matter and antinodes or regions of EM and acoustic radiation. As we have seen acoustic energy is a local phenomenon that does not radiate to the same extent as EM. Thus the various particles known to particle physics begin to condense beginning with the most energetic. This includes the acoustic energy. Quarks begin to appear as the size of the universe grows ever larger. It takes time for the gravitational structure of the early universe to emerge. The nodes of condensing matter eventually form nuclei and hence grow in atomic mass depending on the location with respect to the nodes. At this point, we see the universe is the ultimate eigenvalue problem, a dynamic seething plasma of pseudo matter, vibration, and radiation. As matter at these nodes continues to condense it is pulled ever closer to its centre of mass and the conditions for nuclear reactions like fusion and fission begin to occur at these centres of mass as the heavier elements are pulled towards the centres of mass. According to SFT, the same process causes an internal fission process within the core of the Earth today where uranium has been dragged down to the core; uranium has been collating over billions of years and large pressures have eventually caused the same implosion of an atomic energy reactor to occur at the Earth's core. Thus clumps of matter were created out of the initial acoustic and EM eigenvalues at a precise point in time when the expanding universe was at a critical energy density for those spatial nodes to occur. As the opposite forces of gravitation and expansion were more in balance the dynamism of the universe became more and more like lumps of matter surrounded by various forms of radiation. The various levels of gravitation predicted by SFT took another epoch to emerge beginning with the formation of what became super clusters and eventually resulting in the stars with their planets.

There is another connection to the phonon within cosmology that might well serve to validate SFT at both the strong nuclear and the galactic domains. Before discussing the dynamics of galaxies with respect to SFT, the state of hypothesis within astronomy regarding the observational evidence from our own Milky Way Galaxy is pertinent. Shown in Fig. 4.14 is an up–down motion known to be associated with our own Solar System. Figure 4.15 shows the rotational velocity of a galaxy with respect to radial distance

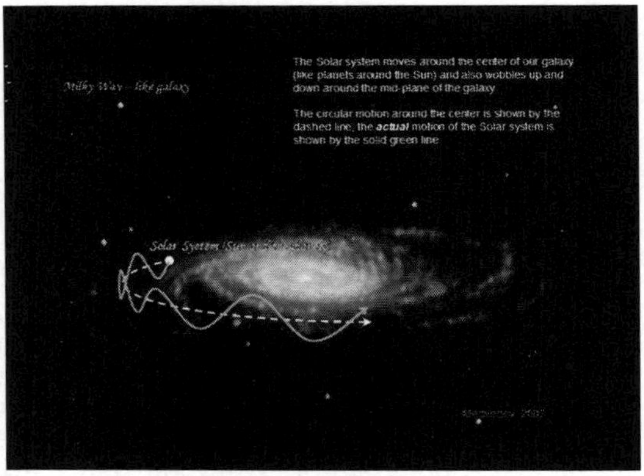

Figure 4.14. Galactic dynamics: typical sun orbiting around the centre of a galaxy. (Source: http://kusmos.phsx.ku.edu/~melott/mmcr.htm) See also Color Insert.

from the centre of the galaxy. As seen it shows that the galaxy appears to move as a solid conglomerate of matter. There have been two lines of hypothesis to explain this form of rotational motion. "Dark matter" can be inferred from the amount of visible matter in structures of galactic size and larger. Dark matter can account for the missing mass in calculations of galactic rotational speeds, the orbital speeds, and gravitational lensing of background objects of galactic clusters, and the heat distribution in galaxies and clusters. Dark matter may play a role in structure formation and galactic

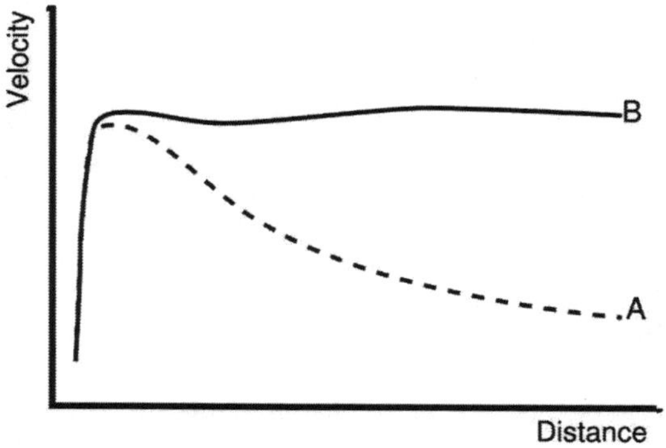

Figure 4.15. Rotational velocity from centre of galaxy. (Credit: Wikimedia.)

evolution. Dark matter is thought to be a form of matter that does not interact electromagnetically. Another hypothesis to explain the observed galactic dynamics is a proposed modification of Newtonian dynamics (MOND). The orbital velocity of circular orbits can be estimated as $v = \sqrt[4]{G M a_0}$ where G is the Newtonian gravitational constant, M is the mass of the star, and a_0. is the orbital distance. According to SFT the solution to this puzzle may be a combination of these two hypotheses.

To understand the process by which the phonon can induce the formation of solid structures at the galactic level we need to survey both the use of sound waves, cymatics, as a modality for structure formation and Pedology, the study of soil including its formation from rock. The reverse process is the formation of solids or conglomerates from particulate matter such as dust and soil. Jenny a pedologist, who termed the name cymatics, performed a series of experiments exposing metallic plates, often circular, covered with sand, water, and other substances to acoustic frequencies (see Figure 4.16; compare Figure 16b and 17). This revealed the remarkable capability of sound and its modes to influence structure in various phases of matter including solid and liquid.

The gravitational structure within galaxies appears to be a differential form of the strong nuclear forces. This is analogous to the discussion in Chapter 2 of the gravitational structure at the solar system level where a dipolar form of EM forces was suggested as the basis

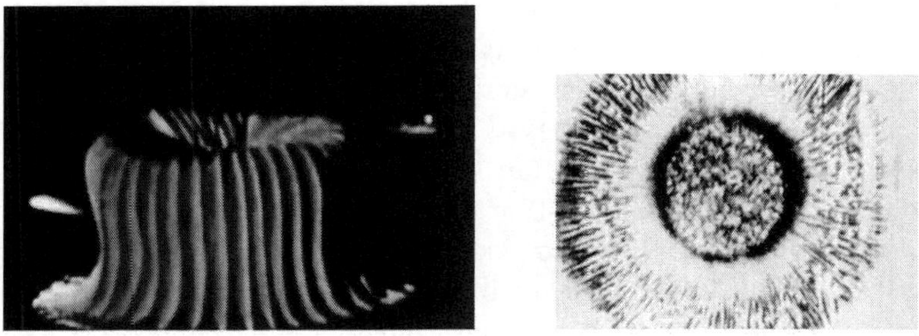

(a) (b)

Figure 4.16. Exposure to sound (a) liquid plastic appears as a cake (b) iron filings as an eye. (Source: http://www.unitedearth.com.au/sound.html) See also Color Insert.

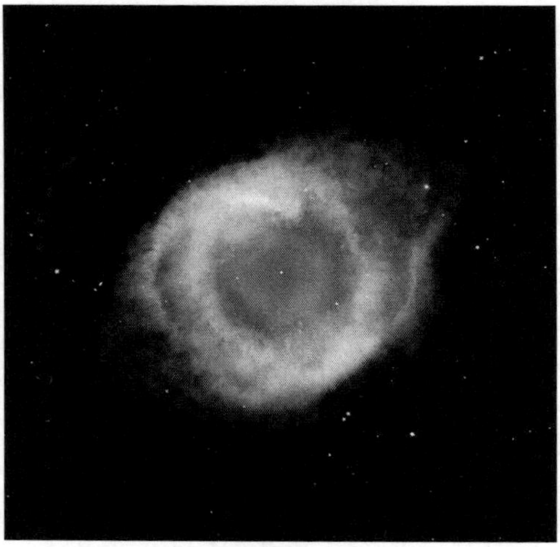

Figure 4.17. Helix nebula, NGC 7293, sometimes known as "The Eye of God." (Credit: Wikimedia.) See also Color Insert.

of the gravitational forces and in Chapter 3 where a tri-spinorial basis was suggested at the galactic level. At the galactic level stars obey a form of gravitation that appears to be somewhat different to that experienced by planets within solar systems of which there may be hundreds of billions in any individual galaxy. Galactic dynamics are known to include oscillations in the (θ, φ) plane indicative of a differential form of the strong nuclear forces. If it was a bi-spinorial form of motion then there would be no phonon involved in the boson-mediated motion as is the case at the solar system level. However, if suns within galaxies also rotate around their own axes in response to galactic forces as well as any planets within its own solar system, then the motion is tri-spinorial and phonon based. Importantly there will a 3-D grid of forces that bind all matter in a galactic structure together like a solid and hence there will be a constant rotational velocity. In retrospect both dark matter and MOND are seen to be in some ways valid but the actual solution may be a combination of these two hypotheses in the form of the tri-field or tri-spinorial equations (2.45) given by SFT.

There is a related study of cyclic patterns of biodiversity over astronomical time periods that have been related to the solar

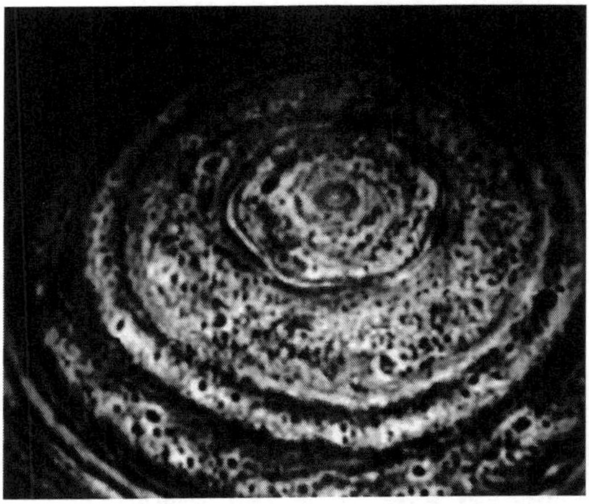

Figure 4.18. Hexagon structure surrounding north pole of Saturn of side about. (Credit: NASA/JPL/University of Arizona.) See also Color Insert.

oscillatory motions in the (θ, φ) plane. At present the cycles of biodiversity are thought to be associated with nearby cosmic ray events. Alternatively, the evolution of biological life forms may be a resonance process related to the energy density levels within the Milky Way Galaxy.

There is another recently discovered phenomenon at our own Solar System level that may well demonstrate the involvement of the structural forming effects of the phonon. A strange unexplained hexagonal structure has been found to encircle Saturn's north pole. This structure is considered due to an acoustic standing wave being generated within Saturn's atmosphere. Each side of the hexagon is of the order 14,000 km (see Figure 4.18).

4.4 Solving the Equations of Electromagnetoacoustics

To the author it appears that science is in a similar situation to the state of research immediately after the EM equations were initially formulated by Maxwell. The four Maxwell's equations are known as Gauss's law, Gauss's law for magnetism, Faraday's law of induction (see Figure 4.19), and Ampère's law with Maxwell's correction in adding displacement currents. To this

Figure 4.19. Induction coil used in 1800s to show Faraday's law of induction $\varepsilon = |\mathrm{d}\phi_B/\mathrm{d}t|$. (Credit: Wikipedia.) See also Color Insert.

mathematical artillery we must add Faraday's concept of "field" that we now understand to be a photon flux. It was also Faraday's discovery of electromagnetic induction in 1831 that inspired Maxwell to search for the unifying theory of electromagnetics. With the theory in place, the technical advances in electromagnetics such as transcontinental communications, telephone, alternating power production, and induction motor were implemented. It has taken another century to add Lorentz's electromagnetic force equation within SFT to solve the atom with its photonic underlay analytically rather than via quantum theory that predicts energy densities or clouds of probability. It appears somewhat ironic that it was Maxwell's work in statistical distributions that eventually led to a method to circumvent the exact solutions that lay hidden until uncovered by SFT. The key to these exact solutions centres about the fact that EM is essentially bi-spinorial.

There have been many studies of phonon-photon interactions. One well-known interaction is the thermal de-excitation of solids or the energy states in semiconductors. When a dense crystal structure is energized via photon excitation, it can either settle back to its previous energy state via photon or phonon release. The phonon mechanism can be an incoherent form of vibration or heat whereas the photon de-excitation is coherent and can be used for communications purposes. Coherent energy exchange between electrons with phonons and photons is an important mechanism in semiconductors. Recently a method was found to transfer encoded information from light in the form of a laser beam to sound and back to light waves again and represents a step towards superfast optical communications. Computers currently use electron-based circuitry while photon-phonon based computing will be a significant leap forward in computing capabilities. Similarly, methods of measuring ultra-short pulses of light are enabling spectroscopy of molecular vibrations, biomedical imaging, electron dynamics, ultra-high precision spectroscopy and optical frequency metrology.

It seems there is a completely new science awaiting experimental exploration. As suggested when discussing how clouds may produce lightning, wind, and thunder via quantum state changes, these effects may be due to quantum states associated with the strong nuclear forces. Till now, the main avenue of experimental study of the quarks and gluons that interact inside protons and neutrons has been particle physics. Yet SFT indicates a system of equations based on tri-spinors including a modified system of Maxwell–Lorentz (ML) equations. These equations indicate a system of quantum states similar to those inside the EM region of the atom but applying to the strong nuclear region. There should be six quantum numbers applying to the nucleus and these correspond to the various "flavours" known to exist. Like the EM forces this corresponds to the equipartition law, there being three degrees of freedom of motion. A nuclear "sonoscope" or stethoscope may enable another way of exploring the strong nuclear quantum states other than via particle physics.

At this embryonic point in the development of a comprehensive tri-spinorial phonon-photon theory, there appear to be two main issues that need to be addressed. These are (1) the creation of a descriptive terminology across phonon-photon phenomena and (2) finding a standard way of expressing electromagnetoacoustic (EMA) fields in mathematical form. Both problems are related to each other and to the very recent status of this line of physics research. The terms acoustic, sonar, and phonon can be married with various effects that relate to heat, stress, strain and elastic deformation. Similar to the term photonics the term phonics or phononics are occasionally used to indicate acoustic or sonar effects including electronic components. The composite term phono-photonics indicates the quantum level field of EMAs. The problems are associated not only with the type of physical phenomenon involved whether stochastic or coherent but also the physical phase of the medium, whether in the solid, gas, or liquid phase, not to mention plasmas or Bose-Einstein condensates. EM effects are treated in general by determining experimental values of the permittivity and permeability of differing materials over several decades of frequency although much work remains at lower levels of interaction. Similar measurement of the speed of sound as it varies over decades of frequency and in different materials is needed. Thus a complex interrelationship is needed across permittivity, permeability *and* the speed of sound. Measurements of EMA phenomena at various ambient temperatures and heights above sea-level are required In the same way that the speed of light is used to define applications to EM as a coupled electric and magnetic effect, we need to understand the way the three fields E-, H-, and A-fields interact parametrically. We probably need to crawl before we walk. Only when we understand electroacoustic and magnetoacoustic effects will we be able to understand the underlying strong nuclear effects.

In the physics of EM phenomena, the same problems of terminology and standardized equations that currently face the science of the phonophotonics have been around for the past 150 years, since Maxwell discovered the 20 equations of EM. Heaviside and independently his peer Hertz succeeded in reducing Maxwell's 20 equations into a smaller set of curl and divergence equations in complex variables. Heaviside coincidentally like his peer Edison was a telegrapher, suffered from poor hearing, and had not been a university trained academic. It did help that his uncle was Charles Wheatstone who invented the Wheatstone bridge, a method of physically comparing lumped resistances. A generation later Einstein introduced the symbolic indices that simultaneously synthesized but obscured our understanding of physics. Dirac another generation later followed this tradition of compacting physical information into a smaller and smaller number of symbols. Currently SFT sees the speed of light not as the relativistic upper limit on speed within physics. There appears to be at least one level of field below the photon that obeys SFT and holds the photon together, similar to the binding fields within the hydrogen atom. In short, EM has undergone some immense changes in conceptual framework over the past 150 years. Looking at the appendix in Jackson's *Classical Electrodynamics* concerning EM units and the various forms of Maxwell's equations provides a historical perspective to this question of how to express the equations of the phonon.

It has been seen how a system of modified ML equations may help clarify nuclear physics. The trouble is we do not currently know of such theoretical equations. The well-used, well-known wave equation of acoustics is a second-order partial differential equation

$$\nabla^2 p - \frac{1}{c_s^2}\frac{\partial^2 p}{\partial t^2} = 0 \tag{4.7}$$

where the speed of sound in air is given by $c_s \approx 3.314 \times 10^2$ m/s, about 10^6 times smaller than the speed of light in free space.

$$c_s = \sqrt{1/\rho_0 \kappa} \tag{4.8}$$

where ρ_0 is the mean density and κ is the adiabatic bulk compressibility. The speed of light c is defined as a fundamental constant (Appendices A and B). The permittivity of free space ε_0 is evaluated via the theoretical expression $\varepsilon_0 = (\mu_0 c^2)^{-1}$, where both c the speed of light and μ_0 are defined to be constants. This wave equation is similar in form to the derived EM wave equations for the E- and H-fields given by equations (1.1) and (1.2)

$$\left(\nabla^2 - \frac{1}{c^2}\frac{\partial^2}{\partial t^2}\right)\vec{E} = 0$$

$$\left(\nabla^2 - \frac{1}{c^2}\frac{\partial^2}{\partial t^2}\right)\vec{B} = 0$$

Does this mean we can use the three wave equations of (4.7) and (1.1) and (1.2) in the three fields to calculate effects? Probably, but only in the "acoustostatic" limit analogous to the uncoupled nature of electro- and magnetostatic effects due to slowly time-varying EM effects at low frequencies. The obvious and best format for defining the interaction appears to be the modified tri-field form of the ML equations given by (2.45). At this point in time, it is not known what mix of parameters if any might sustain valid mathematical solutions. While there is nothing known to impede the existence of valid physical tri-field solutions there is no guarantee that the mathematics of (2.45) will yield physical results. At this stage mathematical research is needed. If valid solutions can be found high-energy physics will no longer be dominated by collision dynamics. The door will be opened to a dynamic form of chemical reaction where collisions become well defined motions as discussed by King and Harrison. In these equations we find the constitutive parameter for the nuclear field is v_n which corresponds somehow to the physical parameters in (4.8). By matching theory, the solutions to (2.45) for experimental spectroscopic measurements of the nuclear frequencies corresponding to (4.8) might enable a complete physical understanding of nuclear processes.

Chapter 5

Self-Field Theory: A Mathematical Model of Physics

The case has been made for self-field theory as a new mathematical description of physics applying across the known frontiers of physics from sub-photon to multiverse. This chapter looks at where current science sits at the start of the 21st century and then looks towards the future of science in the light of self-field theory. In general self-field theory gives mankind a new mathematics that makes current quantum theories outdated, although these theories can be modified to become self-field theory analytic tools to apply to the tasks of the future. Self-field theory is compatible with relativity and in the main with the results of current quantum theories. By using a new bi-spinorial model for relativistic electromagnetic (EM) interactions Heisenberg's uncertainty principle (HUP) is obviated. The seeds of a new analytic era of atomic and nuclear chemistry are visible including the composite photon, phonon, and gluon allowing the quark to be used to model protons and neutrons. New insights of crystal behaviour are apparent including a proposed solution to the double slit phenomenon. EPR (Einstein, Podolsky, Rosen) too becomes less "spooky" and more about coherent atomic structures within crystal lattices bound by photonic bonds. Such

Self-Field Theory: A New Mathematical Description of Physics
By Tony Fleming

www.panstanford.com

bonds may be due to the two new quantum numbers predicted by self-field theory to be associated with the photon in its role as binding agent. These concepts allow further insight into planetary behaviour including earthquakes, plate tectonics, ionospheric layers, and magnetic flips and permit understanding of the way lightning, wind, and thunder are produced by temperature drops within cloud formations. In this era of global warming scientific knowledge will be invaluable. Energy production via fusion engineering is already destined to emerge reducing the imperative for populations to dwell in overpopulated cities. Resonance exposures of the cells of both healthy and unhealthy tissues and organs may emerge to challenge the use of surgery or chemical modes of medical therapy. While the 20th century was often an era of "dirty" technology self-field theory may allow the 21st century to be based on cleaner technologies. A new science that is able to model physics down to the photonic level will be a more people friendly tool for the future where the environment, communities, houses, and people will be protected from hazardous levels of exposure to photonic emissions.

5.1 Repairing Past Inaccuracies

Over almost a century quantum theory has stood at the pinnacle of mankind's mathematical knowledge of physics. Planck's discovery of quantum physics was revolutionary although it was foreshadowed by both Faraday and Fourier more than 50 years earlier. The evolution to quantum mechanics (QM) in 1927 solved the immediate problem of analyzing the atom, bringing about a "golden age" in analytic chemistry that resulted in the eventual discovery of the double helix of DNA. In the longer term QM came with side issues and problems that could not be circumvented until self-field theory (SFT) and its realization that the photon should be modelled using bi-spinorial mathematics. This brings relativity into the model in a different way than originally formulated by Dirac as an added term to the Lagrangian. In this new form of relativity, the added parameter, distance, in SFT overcomes the inaccuracies of HUP. Other than the simplest hydrogen atom model QM, quantum electrodynamics (QED), and quantum chromodynamics (QCD) have been unable to solve the atom deterministically and numerical errors are a fact of life for the analyst. Many-electron and many-atom problems are treated using numerical approximations such as Hartree–Fock. SFT allows deterministic solutions for the eigenvalue problems associated with EM, weak, and strong interactions.

Various disciplines of engineering have been limited by the inherent inaccuracies contained in quantum theory in the 20th century including nuclear, chemical, and electrical engineering. Perhaps science of the past century is best exemplified by the production of the atomic bomb at the end of World War II (WWII). At that time the effects of the weapon were not understood in any detail whatsoever apart from the fact that it was extremely big, bigger than any other bomb in history. It was conjectured that it might even ignite the atmosphere. That Heisenberg and the Nazi scientists who were attempting to match the Allied effort did not develop the weapon was serendipitous. In comparison with such energies, the self-energy of photon is negligible and to say otherwise was at that time ludicrous. But this was indeed throwing out the baby with the bathwater in terms of the mathematical physics contained within the photon. Only in knowing the photon mass can the nuclear motions also be calculated.

Poor air quality was known in the Middle Ages with the use of coal in densely populated cities such as London but became more pronounced with the onset of the industrial revolution in the late 18th century. That urban air quality is very poor in developing countries such as India and China is testament to where these nations sit on the development ladder. Where large populations are being raised out of poverty air quality is unfortunately not seen as a top priority among development aims.

Many domestic, industrial, and agricultural chemicals were developed in the 20th century in the era of quantum theory with its limited accuracy. The developed and undeveloped world continues to suffer chemical pollution of various kinds as the unfortunate cases of ddichlorodiphenyltrichloroethane (DDT) and thalidomide attest. Pharmaceutical side effects are de jour at present. Chemotherapy is based on the principle that cancerous tissues are more sensitive to toxins than healthier tissues. Certainly this can be better than no treatment but this strategy is rapidly becoming not good enough.

Rising levels of male infertility have been linked to water pollution via chemicals that inhibit testosterone including cancer treatments, pharmaceutical treatments, and pesticides used in agriculture make their way into the water system inhibiting fertility across species. Other chemicals mimic estrogens causing male fish to change sex.

Inner city living is another cause of pollution that afflicts all corners of the developed world with rising rates of crime, stress, noise levels, and other forms of urban problems. While urban renewal projects are able to minimize some of these effects, outlying commuters have a need to travel to city centres. Traffic and noise are endemic problems in highly populated cities such as Mexico City and Los Angeles. Acoustic and submarine sonar standards are required in the same way that EM standards have developed since the end of WWII.

Unlike other forms of pollution EM (termed EMR electromagnetic radiation) is internal to the body and mainly invisible; the internal tissues of the body do not in general feel pain as opposed to the skin for instance. The impact of modern technologies on human health such as the mobile phone and transmission towers are in question and depend on the definitions used to define the exposure. The current standard of allowed EM dose is given by a unit called specific absorption rate (SAR) that is measured over a macroscopic cube of tissue (1 cm^3). While SAR is limited, overall effects producing a rise in temperature of 1°C are considered hazardous. The trouble with this measure is twofold. First, only so-called thermal effects are recognized. As Planck's blackbody shows all frequencies are thermal hence the premise is correct, only thermal effects do indeed exist, but the frequencies and effects that do not produce a macroscopic measure of heat cannot arbitrarily be ignored as ineffectual such as those at extremely low frequencies (ELF) or those that produce effects other than a rise in temperature of. Second, cells and other biological systems are only a few micrometre in size, so they cannot produce a 1°C increase in temperature. It seems the goal posts have been adjusted to smear out any so-called non-thermal effect. SFT with its photonic level of interaction stands poised to impact EM standards in terms of allowed electrical and magnetic energy levels over the entire frequency range rather than a temperature rise over a tissue volume.

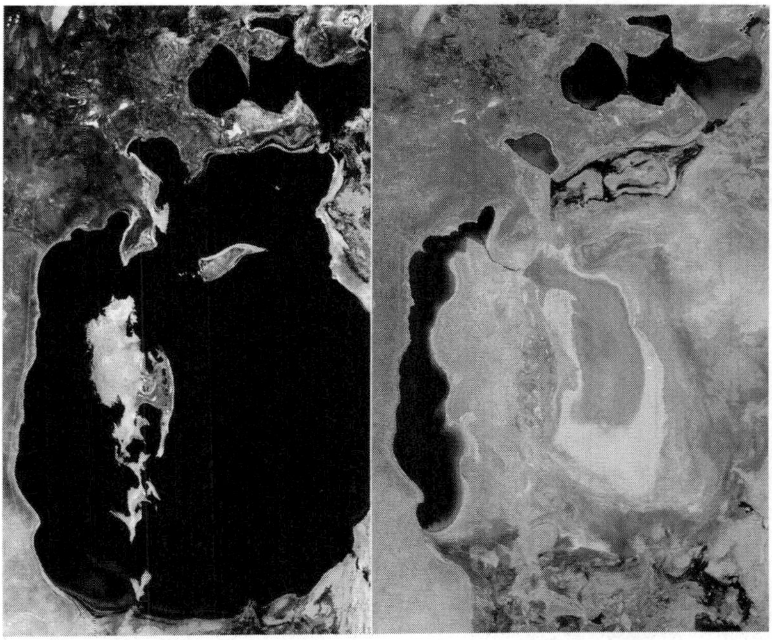

Figure 5.1. Drying of the Aral Sea due to Soviet irrigation diversion. (Source: Wikipedia.) See also Color Insert.

The list goes on. Other forms of pollution such as marine pollution exist. The bleaching of coral reefs threatens areas of multi-diversity such as the Great Barrier Reef. Perhaps the most tragic instance of man's misuse of scientific technology is the drying of the Aral Sea due to misguided Soviet era irrigation projects (Fig. 5.1).

The 20th century may well be seen as the "dirty" century as time goes on. That man came to edge of the abyss and looked over is the reality of the atomic age that followed WWII. Mutually assured destruction or MAD will stand out in human history as the peak of man's insanity as the world went to war for a second time barely catching breath before entering the Cold War. Far from being the "Godsend" it is, nuclear power has been a trial for humankind thus far in its development. SFT promises to revolutionize nuclear analysis by allowing closed form solutions to nuclear problems thus refining the mathematics and hence the engineering.

As we venture into the 21st century both science and technology need to rethink their premises at the most fundamental level. The Copenhagen meeting on climate change in December 2009 was in many ways a test of humankind's resolve to counteract all the negatives listed above. The aim of the meeting was to attempt to lock the world's nations into a cap and trade agreement on carbon emissions. The vast majority of the world's scientists are observing that carbon emissions have reached dangerous levels and that this is the reason for the climate changes measured over the past few decades. While such global warming may well be the result of humankind's own activities since the industrial revolution, sceptics, and a lack of unified resolve at the political level produced a weak outcome at Copenhagen.

While the above does not make good reading for advocates of science, we focus for a minute on what science and technology have achieved over the past century. The landing of man on the moon is testament to the genius of man's technological abilities. The ability to overcome technological hurdles was demonstrated by the Apollo 13 mission that almost ended in disaster. Like the modern Olympic Games, science has become a force for international cooperation. While Los Alamos was directed towards a most terrible weapon the project gave indication of the international effort that may in the future be required for major technologies. This type of international project may well be required into the future as we search for new ways to produce energy and power our transport needs. There are signs for optimism that national interests are evolving into global ambitions of growth and development where each nation brings something to the global marketplace. Cold War adversaries now recognize each other with interconnections growing each day. The telecommunications networks that began with Hertz have resulted in the World Wide Web a tool for spreading information across the globe. It is difficult to overestimate the impact this network will have on global growth and development over the next century. Technological and medical information is rapidly spreading this knowledge to undeveloped parts of the globe. Our great-great grandmothers would not have believed that television would appear in almost every home on Earth. We will no doubt like our great-great grandmothers be amazed at the technologies to be brought into existence by the year 2100. Scientists, like musicians and other cultural performers give example for the politicians to follow. We live between an age of war and an age of clean sustainable development.

Appendix A

Mathematical Preliminaries

A.1 Solving Maxwell's Partial Differential Equations

Since the time of Newton's "fluxional" equations up to present three main techniques have been used to solve ordinary and partial differential equations, the two branches of analytic methods, differential equation methods and integral equation methods, and thirdly the numerical methods. Leibniz and the Bernoulli brothers were among the first to propose differential methods in the late 17th century, with the theory of particular and general solutions emerging by the 19th century along with an understanding of the eigenvalue problem. In the mid-18th century Euler and Lagrange developed a theory based on finding the stationary values of functionals that led to Lagrangian field theory. Fourier and Green implemented integral equation methods early in the 19th century. Solutions of analytic methods are deterministic and closed form rather than the infinite series that are required in numerical methods with their truncation errors. Ancient approximation methods are known from the time the Babylonians first wrote in mathematical notation ca 5000 BC including an estimation of π via an early form of finite elements. Solving the roots of polynomials was an ancient problem that was discovered to be associated with the solutions of differential equations during the 19th century. The development of numerical

Self-Field Theory: A New Mathematical Description of Physics
By Tony Fleming

www.panstanford.com

methods for obtaining these roots including geometric methods and Newton's iterative method had begun during the Middle Ages. Numerical methods and matrix theory grew during the 19th century and were seen as alternatives to the analytic methods. Today various computer-based numerical methods are used as an indispensable adjunct to the kitbag of methods used by engineers and scientists across the globe.

While analytic methods were well known following the advent of quantum theory around 1927, numerical methods grew in step with the acceptance of Heisenberg's uncertainty principle (HUP) and the search for methods of solution of the equations of electromagnetics. The closed-form deterministic methods of analytic methods were precluded by the quantum theories that except for the hydrogen atom failed to solve the quantum equations analytically. It seemed natural that the classical electromagnetic (EM) theory that had failed to analytically solve Maxwell's equations would be replaced by the quantum theories that gave reason, HUP, for the failure. Heisenberg considered uncertainty was a physical block to knowledge of the photon. Along with the use of the EM wave equations various theories of gauge symmetry emerged with quantum theory. As World War II raged across the globe, the invention of the computer led to a widespread implementation of quantum theory with its intensive matrix computations.

Explicit in quantum theory are HUP with its probabilistic nature and a heuristic basis where Planck's "constant" is immersed within a Lagrangian eigenvalue formulation, a wave-like equation. The present-day standard model is also imbedded with HUP and a heuristic eigenvalue basis where tensor mathematics is matched to the experimental evidence from particle physics. For example, the Lagrangian for the spin 1 fields $-\frac{1}{4}\int B_{\mu\nu}B^{\mu\nu} + tr\,W_{\mu\nu}W^{\mu\nu} + tr\,G_{\mu\nu}G^{\mu\nu}$ is a re-normalizable gauge field. While such Lagrangian mathematics has served science over the recent decades time appears ripe for an alternative and conceptually much simpler mathematics that involves Planck's number without heuristic intervention, without explicit need of gauge symmetry, nor HUP.

Nowadays the well-known Maxwell's equations are accepted by students and researchers without question. Yet these four partial differential equations took some decades and much debate to evolve. Maxwell used 20 quaternion equations to describe electromagnetism in his original 1865 paper, which was presented orally to the Royal Society in December of 1864. Maxwell had used a scalar and vector potential as variables in his original equations. Heaviside and Hertz converted these quaternion equations into vector differential equations. Lorentz later applied a symmetric gauge to convert these into the vector equations in the E- and H-field variables we now call Maxwell's equations. There have been many other circuitous roads used along the way. One major byroad involved the two potential wave equations that were later applied to relativistic quantum theory. One motivation for the potential basis was the desire by Maxwell and others to demonstrate and investigate the underlying wave nature of the EM phenomenon. This line of research resulted in the eventual widespread application of radio.

The current situation is that quantum theory solves the wave equations via integral equation methods that are applied as numerical perturbation methods involving infinite terms resulting in probabilistic solution variables known as observables. While gauge theory is mandatory in quantum field theory (QFT) with its potential variables, no such groups are necessary in self-field theory (SFT) as Maxwell's equations (as derived by Lorentz) are used and these gauge conditions are a priori an implicit element of any analytic solution. While QFT uses the second-order wave equations, SFT uses the first-order Maxwell's equations together with the Lorentz force equation to provide analytic closed-form deterministic solutions using differential methods. In Chapter 2 the particular spinor solutions for the E- and H-fields are tested prior to the general eigensolution for the hydrogen atom, forming a classical differential eigensolution method.

SFT does indeed return us to a mathematical age before 1927 by re-examining classical EM theory at the atomic level. However, while the partial differential equations of the classical EM theory do not change the mathematical form of the E- and H-field do. Relativity is achieved not via a Lagrangian term involving spin but by the bi-spinorial motion of the fields. Both E- and H-fields become streams of particles that can be modelled in time evolution via Dirac-delta functions, in some ways similar to QFT. SFT's time-dependent E- and

H-fields are applied as field pairs that differ from the time-static uncoupled Coulomb and Ampere forms of classical EM field theory. The SFT field forms are coupled at the atomic level. In classical EM theory electrostatic and magnetostatic fields are uncoupled forms of the SFT bi-spinorial fields that approximate over macroscopic domains and time-invariant applications to their classical counterparts. Jackson recently investigated the conditions where classical EM is valid and where quantum theory takes over as the domain size shrinks. According to Jackson the mass of photon is central to this issue and as we see in Chapter 2 a non-zero photon mass is associated with SFT. On the other hand QFT demands a zero-mass photon due to gauge symmetry. However, as in the early self-fields of Abraham and Lorentz an analysis of *pairs* of particles within quantum electrodynamics (QED) and *triplets* within quantum chromodynamics (QCD) will enable gauge symmetry to be maintained within QFT. This entails inserting bi-spinorial field forms of SFT within QFT.

It is SFT's bi-spinorial field functions that supply completeness to the over-constrained mathematics of current quantum theory; this bi-spinorial form achieving the relativistic motion of the particles and fields. Stability of the SFT solution is also achieved via the bi-spinorial field that eliminates HUP at the same time in comparison to the asymptotic stability required by QFT. On the other hand QFT can be compared with the numerical method known generally as the integral equation method that includes finite element and boundary element methods, SFT in its present form can be thought of as an analytic finite difference method in comparison to the numerical method known as finite differences. Yet the fractal nature of SFT demonstrates its non-classical nature. If the modern finite element and boundary element methods are applied to QFT with the correctly constrained or complete bi-spinorial field functions a more refined mathematical description of physics beyond the current early implementation of SFT will become available. SFT has thus far provided a complete "Bohr theory" for the atom; it remains to be applied to the composite atomic models described in this book.

Perhaps the most exciting element of SFT is that rather than a binding energy shared between many electrons and protons within an atom, its EM fields apply only to *pairs* of particles, and its strong fields apply only to *triplets*. This extension of current atomic theory to include the binding process along with the concept of orthogonality within space-time may enable the fully analytic closed-form solutions of atoms and molecules.

A.2 Electromagnetic Particle and Field Motions Described via Spinors

The mathematics of classical EM and the QFTs goes back to the mid-19th century. Fields are seen in relativistic dynamics as auxiliary quantities behind particles, and in relativistic field theory as real "objects" able to fill the whole space like a fluid. Scalar, vector, tensor, and spinor fields are defined as mathematical entities.

In QFT spinors are conventionally defined as having constant magnitude, similar to unit vectors. Such spinors are used as bases for Lorentz fields, continuously differentiable objects in 4-D Hilbert space, or pairs of spinors each in 2-D Hilbert space. In particle physics, spinors are two-component complex column vectors of constant magnitude used to represent motions of particles with intrinsic half-integral spin (fermions have spin $\hbar/2, 3\hbar/2, 5\hbar/2, \ldots$). In general, such unit spinors are used as a basis for Hermitian matrices whose determinants are positive definite and whose eigenvalues are all real. The classical EM field equations and the stress-energy tensor are both expressed in terms of unit spinors.

In SFT, physically based variable spinors are used to model the centre-of-motion (COM) particle and field motions. The term spinor represents a space-time vector whose magnitude may vary and rotate extrinsically according to the ML equations. Space-time vectors that rotate (spin) in a single 2-D plane can be expressed in terms of a complex C_2 basis, e.g. $\tilde{r}_o e^{jm\phi}$. The motions of charged particles that perform EM rotations may be termed spinorial. For example, the motion of the electron in the hydrogen atom is described via two spinors since it rotates in two planes simultaneously. At the same time, the photons that comprise the self-field also perform spinorial motions as they move between electron and proton. The fields spiral inwards and outwards towards the proton and the electron respectively, while performing these spinorial motions.

A.3 Centre-of-Motion E- and H-Fields

The spatially ubiquitous nature of the fields currently used within QFT is fundamentally different to the discrete stream-like nature of the COM fields used in SFT. The differences are reminiscent of Planck's initial explanation of black-body radiation in 1900 assuming only certain frequencies were present in the EM oscillators. In classical EM, for instance Coulomb's law, all solid angles are assumed possible regardless of the geometry (Fig. A.1). On the other hand the COM fields have a particulate or quantum nature imbedded in them a priori; part of the dynamic balance applying in each case depends on a resonance condition where the spatial phase length between particles is a fixed integer fraction of a complete wavelength. In EM applications, this fractional phase distance is normally $\pi/2$. One frequency and a small number of beams transit via the entire range of solid angle 4π steradians surrounding a point charge. This is in complete disagreement with current implementation of quantum theories where all charged particles influence all others regardless of angular position. In SFT space-time orthogonality of the various inner and outer shells and the particles within them isolates pairs of electrons and protons, or alternatively paired electroweak electrons with protons (neutrons). This allows SFT solutions to be closed form or analytic rather than the numerical solutions of QM and QFT

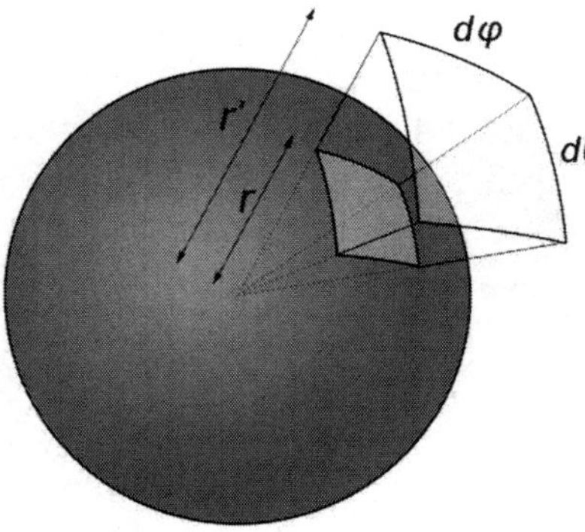

Figure A.1. In SFT a finite number of boson streams transit across a small but finite solid angle as they moves between particles, not over all solid angles. (Credit: Wikipedia.) See also Color Insert.

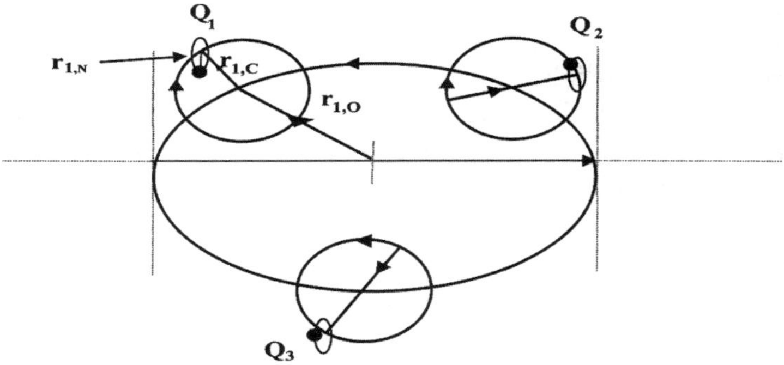

Figure A.2. Proton modelled as three quarks coupled rotating in three orthogonal planes via tri-spinorial gluon fields in the strong nuclear region.

As a tool for mathematically modelling various situations the COM fields can be used widely across physics since there is no difference between particle and field; what is perceived as a particle at one level can be a particle—field interaction at a larger scale from within nuclei to the domain of the universe, as with a fractal. The mathematics of SFT applies to different domains with either a quantum or a continuous, classical-like physics applying to each domain depending on the masses of the sub-particles involved. Various forms of potential and kinetic energies can be linked together using the ML equations and/or modified systems of ML equations inside nuclei where additional spinorial motions and their associated energies are possible. Figure A.2 illustrates quarks moving via three spinors at high energy densities inside protons.

As discussed by Jackson, we must question the assumptions involved in basing classical EM theory, essentially a theory of dynamic charges on Coulomb's law of electrostatics and the Biot-Savart law of magnetic induction a special case of Ampere's law where steady currents apply. It is these two laws that specify the permittivity in free-space as a function of the speed of light and the permeability in free space $\varepsilon_0 = 1/\mu_0 c^2$, where the speed of light is defined as $c \equiv 2.99792458$ m/s exactly and the permeability in free space is defined as $\mu_0 \equiv 4\pi \times 10^{-7}$ H/m exactly. Both laws are based on macroscopic experiments repeated many times but originally conducted in the late 18th to the early 19th centuries. The EM field forms that were later developed are assumed to carry over to atomic, nuclear, and cosmological domains. Further it was natural in the macroscopic domain to assume all actions occurred directly between charges. The fields were also assumed to extend from charged particles over all solid angles to infinity. How the fields imparted forces upon charges was unclear; the fields somehow "mediated" the forces. Bohr applied these field forms to the hydrogen atom with some success but the problem remained until quantum mechanics showed a way forward. One assumption in applying these classical field forms to atoms is that in Coulomb's law they should be static or stationary. Yet atomic charged particles move. Classical fields continuous over all solid angles that were incorporated within QFT may not be correct at the sub-atomic level and other domains as well. This may well be a cause of modelling error including Heisenberg's uncertainty. Jackson also noted the lack of symmetry in the source terms of classical EM theory. The search for a magnetic monopole is related to this perceived asymmetry. In SFT, the H-field is caused by the E-field of a second particle whose photon stream impinges on an observed particle as an H-field. This like the self fields between particles is a mutual phenomenon. Looking at the two particle as an EM binary, there is now symmetry between the two charged particles. The H-field is a matter of reference; E- and H-fields are photon streams relative to the observer's viewpoint.

It is a matter of trial and error to theoretically test other field forms against a wealth of known empirical data. In its usage within the hydrogen atom, the E- and H-fields are taken relative to centres of motion of atomic particles in dynamic equilibrium, and relative to the coordinate system of an isolated stationary hydrogen atom. The finding that these COM forms *can* provide an EM theory consistent with known empirical data for the hydrogen atom is proof of their accuracy at the atomic level. That the E-field forms are inverse square functions of the orbital and cyclotron separations, as in Coulomb's law, and similarly for the H-fields as in the Biot-Savart law, is also confirmed by the analytic solutions derived via EM SFT. The fields are assumed to consist of photon streams and the process of force mediation is similar to kinetic models of gases where numbers of collisions occur over the walls of a container thus imparting pressure over the area of the wall. Photons are considered

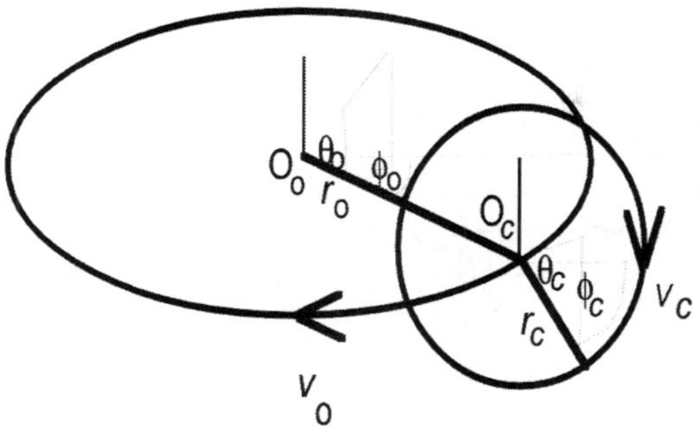

Figure A.3. Orbital and cyclotron radii and velocities of particle showing coupled spherical coordinate systems oriented at the electrical centre of motion (centre of mass). The coordinate systems of the orbital and cyclotron motions are coupled via the EM field equations. Only four of the six spatial coordinates are independent, as in space-time.

to "collide" in some coherent way with the charged particles thus imparting impulses that averaged over time become the fields and forces at the atomic level given by Maxwell's equation. Spatial and temporal averaging is important in the fractal mathematics of SFT.

In the hydrogen atom, the essential physics is the spinorial motion of the photons that move between the charged particles that also perform spinorial motions in concert with the photons. These photonic motions yield both an axially directed force and an axisymmetric or rotational force on the particles when averaged over statistical mechanics numbers of impulses, i.e. the E- and H-fields at atomic levels of interaction.

To mathematically describe the physics behind it, we choose the four independent coordinates to form the two spinors (r_0, ω_0, r_c, ω_c) as shown in Fig. A.3 that are related to the four coupled spherical coordinates (r_0, ϕ_0, r_c, ϕ_c). There is a simple mathematical relationship between these orbital and cyclotron spinors since the E-fields rotate clockwise while the H-fields rotate anti-clockwise. Also, since the mathematics turns out to be an eigenvalue problem, the rotation speeds of the modes are related by integer multiples of the principal mode. We choose to describe the coupling between the rotations via imaginary numbers where the phase shift is $\pi/2$.

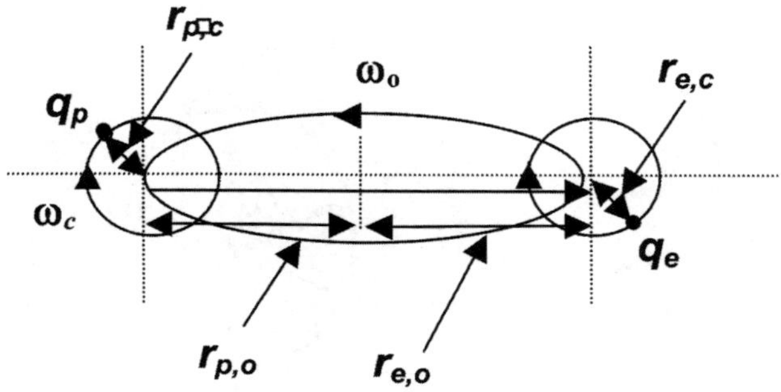

Figure A.4. Geometry for determining the centre-of-motion E- and H-fields. In this diagram both particles have equal magnitudes of mass and charge.

This can be thought of as a phase shift between the spinors that in imaginary numbers becomes simply j.

Figure A.4 illustrates the centres of motion of the fields acting between two charged particles, q_e and q_p of equal mass $m_e = m_p$ and opposite charge $q_e = -q_p$. In this symmetric case, the motions for the orbital and cyclotron radii of both particles are assumed to be circular. In general the relative time-varying orbital radial distance of q_e is given by the spinor $\tilde{r}_{e,o}e^{j\omega_{e,o}t}$, and the cyclotron radial distance by $\tilde{r}_{e,c}e^{-j\omega_{e,c}t}$ while the relative time-varying orbital radial distance of q_p is given by the spinor $\tilde{r}_{p,o}e^{j\omega_{p,o}t}$, and the cyclotron radial distance by $\tilde{r}_{p,c}e^{-j\omega_{p,c}t}$. The overall orbital and cyclotron separations can be written as the following:

$$\tilde{r}_{ep,o} = \tilde{r}_{e,o}e^{j\omega_{e,o}t} + \tilde{r}_{p,o}e^{j\omega_{p,o}t} \tag{A.1}$$

$$\tilde{r}_{ep,c} = \tilde{r}_{e,c}e^{-j\omega_{e,c}t} + \tilde{r}_{p,c}e^{-j\omega_{p,c}t} \tag{A.2}$$

Generally the orbital and cyclotron spinors may be fixed or varying in frequency, in keeping with the need that the total system self-energy remains constant at all times unless external factors intervene. Where the particles are in dynamic balance in a homogeneous and isotropic medium the electron and proton orbital and cyclotron frequencies are equal, $\omega_{e,o} = \omega_{p,o} = \omega_o$ and $\omega_{e,c} = \omega_{p,c} = \omega_c$.

$$\tilde{r}_{ep,o} = \left(\tilde{r}_{e,o} + \tilde{r}_{p,o}\right) e^{j\omega_o t} \tag{A.3}$$

$$\tilde{r}_{ep,c} = \left(\tilde{r}_{e,c} + \tilde{r}_{p,c}\right) e^{-j\omega_c t} \tag{A.4}$$

If needed there is a swapping between the electric and magnetic energies of the particles; for example, when a particle slows down in an elliptic orbital path, its cyclotron motion speeds up, thus the particle energy is fixed.[1] Figure A.4 shows the two coupled spherical coordinate systems that define the orbital and cyclotron distances.

The energy density of the region is controlled by the constitutive parameters ε and μ. In generalized media the permittivity and permeability are related to those in free space $\varepsilon = \varepsilon_r \varepsilon_0$, where ε_r is the relative permittivity and $\mu = \mu_r \mu_0$, where μ_r is the relative permeability. In various media such as metals, dielectrics, or biological tissue, the relative permittivity and permeability can vary by several orders of magnitude and hence the kinetic and potential energies, the forces, fields and energy densities also vary considerably. Mathematically, EM applications have six degrees of freedom per sub-atomic particle: the orbital and cyclotron velocities $\tilde{v}_{e,o} = j\omega_e \tilde{r}_{e,o} e^{j\omega_o t}$, $\tilde{v}_{e,c} = -j\omega_c \tilde{r}_{e,c} e^{-j\omega_c t}$, and the constitutive parameters. In regions of high energy density such as the nucleus, the motions of particles and fields consist of three orthogonal spinors and hence each particle possesses extra kinetic and potential energies than in EM interactions; overall in such regions of high energy density there are eight degrees of freedom, three spinors and the constitutive parameters.

In the case of the proton and electron in the hydrogen atom, where $m_p \gg m_e$, $q_e = -q$, and $q_p = q$, the principal orbital motions are at balance again circular and that is the case of interest in this report. The relative positions of the proton and electron define their E-fields, while their relative positions and velocities define their H-fields. These relative motions also define the proton and electron centres of motion. The forces acting on the electron $\tilde{F}_{E,e}$ and proton $\tilde{F}_{E,p}$ due to the mutual E-field acting between the two orbiting particles are attractive and are written

$$m_e \frac{d^2(\tilde{r}_{e,o})}{dt^2} = -q\tilde{E} \tag{A.5}$$

[1] It is possible to study planets with large eccentricities as a check of energy conservation to see if EM SFT applies to solar systems where dipole moments may apply instead of charges.

$$m_p \frac{d^2(\tilde{r}_{p,o})}{dt^2} = q\tilde{E} \tag{A.6}$$

$$\tilde{F}_{E,e} = \frac{-q^2}{4\pi\varepsilon_0} \frac{\hat{r}_{ep,o}}{\left|r^2_{ep,o}\right|} e^{j\omega_o t} \tag{A.7}$$

$$\tilde{F}_{E,p} = \frac{-q^2}{4\pi\varepsilon_0} \frac{\hat{r}_{pe,o}}{\left|r^2_{pe,o}\right|} e^{j\omega_o t} \tag{A.8}$$

where $\tilde{r}_{ep,o} = -\tilde{r}_{pe,o}$ is the orbital separation vectors and $\hat{r}_{ep,o} = \hat{r}_{pe,o}$ the unit orbital separation vectors. The forces acting on the electron $\tilde{F}_{H,e}$ and proton $\tilde{F}_{H,p}$ due to the mutual H-field acting between the two rotating particles are repulsive. Here, the constitutive parameters are assumed to be dependent on the speed of the particle $\varepsilon_0\mu_0 = 1/(\omega_o^2 r_o^2) = 1/(\omega_c^2 r_c^2)$, the forces can be written as

$$m_e \frac{d^2(\tilde{r}_{e,c})}{dt^2} = -qv_{e,c} \times B \tag{A.9}$$

$$m_p \frac{d^2(\tilde{r}_{p,c})}{dt^2} = qv_{p,c} \times B \tag{A.10}$$

$$\tilde{F}_{H,e} = \frac{q^2\omega_c r_{e,c}\mu_0}{4\pi} \frac{\hat{r}_{ep,c}}{\left|r_{ep,c}\right|} e^{-j\omega_c t} = \frac{q^2}{4\pi\varepsilon_0} \frac{\hat{r}_{ep,c}}{\left|r_{e,c} r_{ep,c}\right|} e^{-j\omega_c t} \tag{A.11}$$

$$\tilde{F}_{H,p} = \frac{q^2\omega_c r_{p,c}\mu_0}{4\pi} \frac{\hat{r}_{pe,c}}{\left|r_{pe,c}\right|} e^{-j\omega_c t} = \frac{q^2}{4\pi\varepsilon_0} \frac{\hat{r}_{pe,c}}{\left|r_{p,c} r_{pe,c}\right|} e^{-j\omega_c t} \tag{A.12}$$

where $\tilde{r}_{ep,c} = -\tilde{r}_{pe,c}$ are radial cyclotron separation vectors, and $\hat{r}_{ep,c} = -\hat{r}_{pe,c}$ unit radial cyclotron separation vectors between charges.

Where the "infinite mass" proton approximation is assumed the E- and H-field vectors for the electron can be written in terms of orbital and cyclotron spinors (Fig. A.5)

$$\tilde{F}_{E,e} = \frac{-q^2}{4\pi\varepsilon_0} \frac{\hat{r}_{ep,o}}{\left|r^3_{ep,o}\right|} \sigma_o \text{ where } \sigma_o(r_{ep,o}, \omega_o) = r_{ep,o} e^{j\phi_o} \tag{A.13}$$

$$\tilde{F}_{H,e} \cong \frac{q^2}{4\pi\varepsilon_0} \frac{\hat{r}_{ep,c}}{\left|r^3_{ep,c}\right|} \sigma_c \text{ where } \sigma_c(r_{ep,c}, \omega_c) = r_{ep,c} e^{-j\phi_c} \tag{A.14}$$

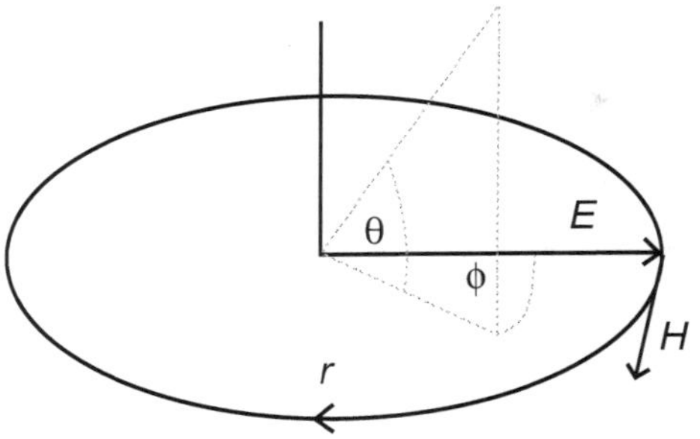

Figure A.5. The hydrogen atom modelled with "infinite mass" proton.

Here the parametric functions of time and speed $\phi_0 = \omega_0 t$ and $\phi_c = \omega_c t$ are written explicitly as orbital and cyclotron phase angles. The radial separation of the charges in this approximate case can be obtained $\tilde{r}_{ep} = \tilde{r}_{ep,o} + \tilde{r}_{ep,c}$. The two components are not generally orthogonal to each other; thus the separation varies with time and is a function of both orbital and cyclotron angular frequencies. The forces can be seen to be centre seeking (centripetal) and centre repelling (centrifugal) and as such form a balanced system in dynamic equilibrium. The overall motion of the hydrogen atom is a combination of all the electric and magnetic forces, a time-varying function of the orbital and cyclotron spinors. By representing the cyclotron phase angle as $\phi_c = -\omega_c t$ and the separations as $r_o = r_{ep,o}$ and $r_c = r_{ep,c}$, the spinors and fields can be written in terms of positive exponentials as

$$\sigma_o(r_o, \omega_o) = r_o e^{e^{j\omega_o t}} \tag{A.15}$$

$$\sigma_c(r_c, \omega_c) = r_c e^{j\omega_c t} \tag{A.16}$$

In the case of the principal mode where one period occurs within both the orbital and cyclotron motions, the phase angles are the negative of each other $\phi_c = -\phi_o$. Within this coordinate transformation, each time we either differentiate in time or space we must be careful to ensure the sign of our magnitudes are correct. While the fields on the electron $\tilde{E}_e$ and $\tilde{H}_e$ are orthogonal at all times, the forces $\vec{F}_{E,e}$ and $\vec{F}_{H,e}$ are not, although they are orthogonal to the velocity of each particle; similar is the case for the proton.

A.4 Imaginary Numbers in 4-D Coordinate Systems

Oftentimes in EM formulations, complex numbers are used from the start, a theoretical procedure is followed, at the end the real and imaginary parts are collected and the imaginary parts are discarded. In this report a physical significance is meant for both real and imaginary parts of the formulation at all times. This said, imaginary numbers are treated no differently within this report to their common usage except that the concept is generalized to the 4-D coordinate system involving the two coupled spinors as illustrated in Fig. A.3. They are "real" in the sense that they are orthogonal entities to mathematically real numbers. Wherever the symbol j appears in 3-D coordinate systems it can be substituted by a real component in an orthogonal direction. We may thus think of it as an "orthogonality operator." Defined in terms of imaginary numbers, 2-D spinors are rotations moving in planes that include axes orthogonal to a stated direction, e.g. (A.3) and (A.4), where time is a parameter of rotation. The interpretation is consistent with the conventional cyclic ordering of coordinate systems, e.g. Cartesian (x, y, z), spherical (r, θ, ϕ), and cylindrical (ρ, ϕ, z). Thus, $j\hat{x} = \hat{y}$. A positive imaginary number is a distance in a positive orthogonal direction and similarly for a negative imaginary number; thus $-j\hat{\theta} = -\hat{\phi}$. It should be noted that an alternative but more complicated notation for imaginary numbers (j, k, l) in 3-D might have been used in this report, whereas the sole symbol j is used to denote orthogonality to any direction as is its common usage in mathematics. The interpretation to orthogonality also holds for the parametric time operator and its differentials. There is implicit in electromagnetics a cycle of four phase shifts of $\pi/2$ involved in the physical directions of differentials whether spatial or temporal. For the strong nuclear forces where three spinors are involved this analytic cycle is longer being six phase shifts of $\pi/3$.

A.5 The Condition for Zero Nett Radiation by Dipole Antennas

The formulation via potential functions of dipole antennas by Hertz in 1889 demonstrated theoretically the existence of an EM wave in

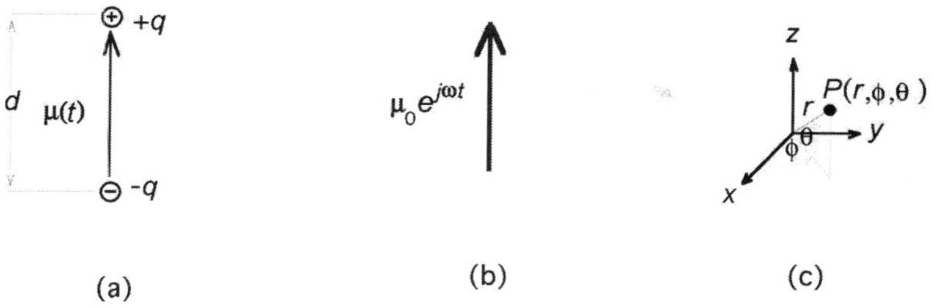

Figure A.6. Dipole radiation: (a) two charged particles forming a time-harmonic electric dipole illustrated in (b) by a dipole moment rotating in a vertical plane, corresponding to the θ direction in the spherical coordinate system (r, θ, ϕ) defined in conventional fashion with respect to the (x, y, z) Cartesian coordinate system shown in (c).

the far field of the dipole (Fig. A.6). Hertz used a loop in the far-field of a spark-driven centre-fed wire dipole to detect an EM wave emitted by an antenna. This theoretical and experimental research was another impetus behind the ongoing revolution in telegraphic and radio communications that was then taking place. Maxwell had theoretically discovered EM waves but physically radiating them was another matter. Hertz discovered the wire antenna that Marconi, Tesla, and others used to transmit Morse code and voice across the Atlantic. This allowed mankind to communicate across oceans and national borders. Although one of first uses was in war, radio has become a dynamic tool for international peace and cooperation. The recent advent of personal mobile phones and computers has enabled an enormous amount of international communications traffic including the Internet that allows the general public to communicate person to person on a global level.

According to the following analysis based on Hertz's original work it is possible via a two-wire antenna combination called a cross-dipole to attain a condition of zero nett radiation (ZNR). ZNR occurs if the system oscillates at a certain resonant frequency and provided the two arms are fed at $\pi/2$ phase difference relative to each other. Thus an open antenna can be closed to the far field.

This not only opens the door to a new form of near-field communications system, but it also allows us to see how at the elementary charge level the atom achieves dynamic equilibrium.

Referred to as an EM self-field binary, two charges can oscillate around each other in their near-fields to induce mutual rotational motions without radiation, a self-field system between EM charges. Following the discussion of Von Hippel who noted that an electric dipole moment written as a complex exponential is equivalent to two linear dipoles, the rotating moments can be written as

$$\mu_\theta(t) = qd_\theta e^{j\omega_\theta t} \tag{A.16a}$$

$$\mu_\phi(t) = qd_\phi e^{j\omega_\phi t} \tag{A.16b}$$

where $$\mu_\theta(\mathrm{t}) = \mathrm{qd}_\theta(\vec{\mathrm{x}} \cos\omega_\theta \mathrm{t} + \vec{\mathrm{z}} \sin\omega_\theta \mathrm{t}) \tag{A.16c}$$

and $$\mu_\phi(\mathrm{t}) = \mathrm{qd}_\phi(\vec{x} \cos\omega_\phi \mathrm{t} + \vec{\mathrm{z}} \sin\omega_\phi \mathrm{t}) \tag{A.16d}$$

and the subscripts θ and ϕ indicate from which dipole the variables are derived. The fields due to these dipole moments can be obtained via the Hertzian potentials. Hertz used these potentials to calculate the radiation flow from a dipole antenna in which only the vertical component of a single Hertzian potential is necessary since the charges oscillate along the z-directed dipole. The use of a single potential for the dipole antenna can be thought of as a rotation in time. While we can continue to use time exponentials to analyze the motion of charged atomic particles, the formulation can be considered to produce a snapshot in time of the radiation pattern that rotates in the far field in both the θ and ϕ planes. The potentials are defined in terms of the dipole moments given by (A.16).

$$O_\theta = \frac{I_\theta d_\theta}{4\pi r_\theta} = \frac{j\omega_\theta}{4\pi r_\theta}\mu_{\theta 0} \exp\{j\omega_\theta(t - r_\theta/v_\theta)\} \tag{A.17a}$$

$$O_\phi = \frac{I_\phi d_\phi}{4\pi r_\phi} = \frac{j\omega_\phi}{4\pi r_\phi}\mu_{\phi 0} \exp\{j\omega_\phi(t - r_\phi/v_\phi)\} \tag{A.17b}$$

The subscripts θ and ϕ again indicate the dipole from which the variables are derived. $\mu_{\theta 0}$ and $\mu_{\phi 0}$ are two orthogonal vectors directed along the z- and x-axes formed by the moments at $t = 0$, while

$$I_\theta = \frac{dq_\theta}{dt} = \frac{1}{d}\frac{d\mu_\theta}{dt}$$

and

$$I_\phi = \frac{dq_\phi}{dt} = \frac{1}{d}\frac{d\mu_\phi}{dt}$$

The system of two rotating dipole moments can be thought of as resulting from a cross-dipole structure having both a vertical arm and a horizontal arm. The B-fields can be obtained as the curl of these Hertzian potentials, and the time derivative of the E-fields is given as usual in terms of the curl of the B-fields as given by Maxwell's equations:

$$B_\theta = \nabla \times O_\theta \tag{A.18a}$$

$$B_\phi = \nabla \times O_\phi \tag{A.18b}$$

$$B_\theta = \frac{1}{4\pi\mu_0 r_\theta^2}\frac{d\,|\mu_\theta|}{dt}\frac{\overleftarrow{d}_\theta \times \overleftarrow{r}_\theta}{d_\theta} \tag{A.18c}$$

$$B_\phi = \frac{1}{4\pi\mu_0 r_\phi^2}\frac{d\,|\mu_\phi|}{dt}\frac{\overleftarrow{d}_\phi \times \overleftarrow{r}_\phi}{d_\phi} \tag{A.18d}$$

Examining the fields of the dipole rotating in the θ plane at a point distance r_θ around the plane θg 0, its radiation components and energy density can be obtained:

$$E_{\theta_\theta} = \frac{d_\theta^2\,|\mu_\theta|}{dt^2}\frac{\sin\theta}{4\pi\varepsilon_0 r_\theta v_\theta^2}e^{j\omega_\theta \frac{r_\theta}{v_\theta}} \tag{A.19a}$$

$$B_{\theta_\phi} = \frac{d_\theta^2\,|\mu_\theta|}{dt^2}\frac{\sin\theta}{4\pi\mu_0 r_\theta v_\theta}e^{j\omega_\theta \frac{r_\theta}{v_\theta}} \tag{A.19b}$$

$$\begin{aligned} dU_\theta &= \rho_\theta dV = \frac{1}{2}(\varepsilon_0 E_{\theta_{\theta'}}^2 + \frac{1}{\mu_0}B_{\theta_{\phi'}}^2) \\ &= \frac{\mu_0\omega_\theta^4\mu_\theta^2(t)}{(4\pi)^2 r_\theta^2 v_\theta^2}\sin^2\theta dV \end{aligned} \tag{A.20a}$$

Similarly, examining fields of the dipole rotating in the ϕ plane at a height r_ϕ around the plane $\phi = 0$, its radiation components and energy density can also be obtained:[2]

$$E_{\phi_{\theta'}} = -\frac{d_\phi^2\,|\mu_\phi|}{dt^2}\frac{\cos\phi}{4\pi\varepsilon_0 r_\phi v_\phi^2}e^{j\omega_\phi \frac{r_\phi}{v_\phi}} \tag{A.19c}$$

---

[2]The distances r_θ and r_ϕ are considered to be much larger than the dipole lengths d_θ and d_ϕ.

$$B_{\phi_{\phi'}} = -\frac{d_\phi^2\,|\mu_\phi|}{dt^2}\frac{\cos\phi}{4\pi\mu_0 r_\phi v_\phi}e^{j\omega_\phi\frac{r_\phi}{v_\phi}} \tag{A.19d}$$

$$\begin{aligned} dU_\phi &= \rho_\phi dV = \frac{1}{2}(\varepsilon_0 E_{\phi_{\theta'}}^2 + \frac{1}{\mu_0}B_{\phi_{\phi'}}^2) \\ &= \frac{\mu_0\omega_\phi^4\mu_\phi^2(t)}{(4\pi)^2 r_\phi^2 v_\phi^2}\cos^2\phi dV \end{aligned} \tag{A.20b}$$

The prime superscripts of the fields refers to a rotated frame of reference in which the radiation components, θ' and ϕ', represent the fields along the meridian (broadside) plane of the dipole moment rotating in ϕ. Apart from the geometric factors in θ and ϕ (effectively a phase shift), the forms of the energy densities (A.20a) and (A.20b) are identical. This is due to the symmetry in obtaining the θ and ϕ components of the curl operator in cylindrical coordinates. Importantly, the flow of energy is now bi-directional: energy associated with the dipole rotating in θ flows outwards through its volume element, while the energy associated with the dipole rotating in ϕ flows inwards. Apart from the geometric factor, what conditions upon the positions and motions of the dipoles can we deduce for ZNR? Equating the two energy densities to provide ZNR

$$\frac{\omega_\theta^4}{r_\theta^2 v_\theta^2} = \frac{-\omega_\phi^4}{r_\phi^2 v_\phi^2} \tag{A.21}$$

If $v_\theta = \omega_\theta r_\theta$ and $v_\phi = \omega_\phi r_\phi$, we may conclude that where $\omega_\theta^2 = -\omega_\phi^2$, there will be no radiated field if between the two components there is a phase factor of $\pi/2$ between their phases, specifically for the case where $\omega_\theta = \pi/2 - \omega_\phi$. By choosing this particular spin relationship, the spatial, temporal, and geometric factors are seen to counteract each other. These ZNR constraints correspond to the spherical orbitals governed by De Broglie's equation $n\lambda = 2\pi r_i$ where $r_i = n^2 r_B$ for $n = 1, 2, 3, \ldots$ where r_B is the Bohr radius corresponding to the principal orbit of the hydrogen atom; elliptic orbits are also possible. The mathematics above treats the fields as classical covering all solid angles, yet the SFT given in Chapter 2 and 3 concerns photons that are streams of field particles of mass and energy.

A.6 The Condition for Zero Nett Radiation in Atoms

Throughout the history of QM up to the present, there has been puzzlement as to how the electrons inside an atom orbit around the nucleus without spiraling inwards to nuclear oblivion, how they remain in dynamic equilibrium without radiating energy. QM evolved from the classical electron of Bohr's model of a specific quantized orbit to a probabilistic quantum mechanical orbital. Within each QM orbital, the electron cloud is more dispersed than its classical counterpart since orbits are considered to incorporate shifts between orbitals that are accompanied by absorption and emission of energy. The problem with the classical model concerns conversion of electrodynamic energy via kinetic energy into radiation and the subsequent spiraling of the orbiting electron into the nucleus as it consumes its own energy. In Bohr theory this is related to taking only one kind of energy, the electrodynamic energy and leaving the magnetodynamic energy aside. A similar error occurs within quantum theory.

At the same time the problem with the current QM solution is that it does not simplify the solution by incorporating the orthogonality of the individual resonances associated within each EM binary in a particular atomic structure. For each mode of the atom, an electron and an associated proton exist in isolated and balanced motion; an individual EM binary does not interfere with any other EM binary within a multi-shell atomic structure. In SFT terms, except for the outer shell electron and its corresponding outer shell proton, there is only one pair of photon streams connecting any particular electron and its associated proton. In the case of the outer shell, there can be several pairs of streams connecting between atoms in molecules and crystals and other arrayed structures including Bose – Einstein condensates. Apart from the outer shell, only pairs of particles can move as EM binaries inside EM regions.

In the outer shells, the electrons and their associated protons must move in a coherent manner with respect to the other members of the array. This coherence across an atomic array may be the reason for the enigmatic double-slit experiments whereby it has been shown using various methods that a picture of diffraction is formed no matter how slowly the photon stream is made. Even if the stream is reduced to a single photon a diffraction picture emerges. If the input stream is stochastic, and therefore unrelated, to the rotations of the electrons and protons in the material from which the slits are removed then a stochastic picture will build up. This can happen only if the atoms are moving coherently; all outer shell electrons are in phase with each other as are the outer shell protons.

Atoms appear to have been composed in a layered pair form with respect to energy as the universe evolved from the original Big Bang. The energy density within the embryonic universe started to diminish as it expanded outwards. Again within the process of galaxy and star formation nucleo-synthesis of the elements occurred beginning with the lightest and evolving through to the heaviest. At times pairs of electrons and protons were made available to continue the evolutionary synthesis upwards in atomic weight of the various elements. Perhaps this was in part a matter of chance of collision versus evolutionary time. It included the process by which acoustic and EM multi-mode resonances formed clumps of matter inside the early universe. Only if the energy density is increased above that within the various atomic shells does the process reverse. In terms of the Lagrangian energy density associated with the EM binary, any two mass-charges of either sign may be stable if they have a phase difference of π between them. The stability point is at the centre of mass and motion with the two charges moving around this stationary point. Remembering the photon within SFT is dipolar in structure, in SFT there are two photons of non-zero mass within an atom with a phase factor of $\pi/2$ between them. SFT concerns two masses and two photons rather than one mass and one photon of QFT, or one mass and one field as with the electron of Abraham and Lorentz.

Landau and Lifshitz pointed out that charges in uniform motion do not radiate, only accelerated charges radiate; a particle that changes its velocity including the electron shown in Fig. A.7 emits energy. Implicit is the notion that curved motion is acceleration. Physicists have long argued that this effect is the reason why electromagnetics could not be the basis of atomic theory.

However, *pairs* of charges that move along curved or curvilinear paths "in sympathy" with one another can move with ZNR if their radiated fields balance each other. Where a near-field standing wave is established between pairs of charges, radiated energy may be

Figure A.7. Electron spiraling inwards due to an electromagnetic field as the electron performs curved motion it radiates which alters its kinetic and potential energies.

mutually balanced between the charges. Take the motions of an electron and a proton in an atomic orbit under the action of their spherical self-field forces. There is a balance of energy over each rotation. The definite integral of electric work over a curve $W = q\int_a^b E\,dr$. Thus the integral of electric and magnetic force times distance must be unchanged over 2π radians in either ϕ or θ directions for stability to be maintained. The electron and proton of the hydrogen atom are a binary system (Fig. A.8). There are E- and H-forces induced on each other by their respective positions and velocities. These forces have forms like Coulomb's law, and the Biot–Savart law but are time-varying due to their relativistic accelerated motions. Bi-spinorial SFT field forms suitable for electro- and magnetodynamic motions must be used; the two particles form an atomic EM binary system.

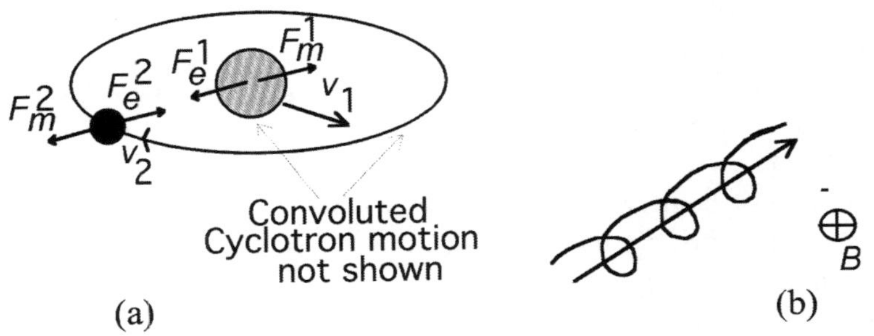

Figure A.8. (a) Atomic EM binary: both particles exert E- and B-fields upon each other. (b) Particle performing a helical motion (cyclotron) due to exposure by a B-field.

Table A.1 illustrates the types of radiation performed by EM particles and gives the point of view of rotation and translation, or orbit and cyclotron, depending on the symmetries involved.

Simple uniform E-fields operate in a translational sense, while simple uniform B-fields in a rotational sense. Leaving aside more complex geometries; electric particles do no work in a B-field, and symmetrically, magnetic particles no work in an E-field. Electric particles can do work in E-fields, and likewise magnetic particles in B-fields. This work may be radiated. Table A.1 shows rotations and translations are associated with magnetic and electric particles, including the undiscovered magnetic monopole. The model of the atom given by SFT is more complex than these simple classical forms. In SFT we find that the photon is also involved inside the atom and collides elastically with both electron and proton as a means of binding the EM binary pair together. There is a range of stability within which the atom is robust to bumps and collisions without becoming unstable. The motions are more or less piecewise linear within a range of collisions that do not knock it out of rotational or cyclotron orbit. This entails a fractal vision of the photon that is essentially dipolar and is attracted towards a dipole moment of E- or H-form as it collides and transits between electron and proton.

A.7 Uncertainty

In Chapter 3, the photon was modelled as two point charges of opposite charge, similar to the hydrogen atom, but of equal mass. The eigenstructure is now continuous in that there is a resonance condition at every energy state of the photon in agreement with Einstein's theory of the photoelectric effect proposed in 1905 and the equation relating photon energy to its frequency $E = h\nu$. If a dipolar photon composed of two mass-charges moves past an observation point, what is the resulting effect? Assume the following EM binary: two oppositely charged particles of equal mass move around their centre of rotation. They move at constant translational speed past point A rotating in the $z - x$ plane (Fig. A.9).

There is a correlation between the theory behind NZR derived previously in this appendix and the uncertainty principle first

Table A.1. Field forms and their ability to do work

Field	Particle	Force	Velocity	Motion	Form of Work	Radiation
Constant E_x-field	Electric monopole q_e	$q_e E_x$	$v_x = 0$	Accelerated translation	$\frac{1}{2}mv_x^2$	Yes
Variable E_x-field	Electric dipole p	$p\frac{\partial E}{\partial x}$	$v_x \neq 0$	Accelerated translation	$\frac{1}{2}mv_x^2$	Yes
Constant H_x-field	Magnetic monopole q_m	$q_m B$ (by symmetry)	$v_x \neq 0$	Accelerated translation	$\frac{1}{2}mv_x^2$	q_m probably does not exist
Variable H_x-field	Magnetic dipole μ	$\mu\frac{\partial B}{\partial x}$	$v_y, v_z \neq 0$	Accelerated rotation	$\frac{1}{2}mv_y^2 + \frac{1}{2}mv_z^2$	Yes
Spherical H-field	Electric monopole q	$q_e v \times B$	$v_\phi, v_\theta \neq 0$	Rotational velocity	Changes in $\frac{1}{2}mv_{cyclotron}^2$ within cycle	Within cycle yes but no, ZNR over 2π cycle
Spherical E-field	Electric monopole q	$q_e E$	$v_\phi, v_\theta \neq 0$	Rotational velocity	Changes in $\frac{1}{2}mv_{orbital}^2$ within cycle	Within cycle yes but no, ZNR over 2π cycle
EM Binary atomic level	Two electric monopoles q_e	$q_e E + q_e v \times B$		Convolutions between translations/ rotations	No change in $\frac{1}{2}mv_{orbital}^2 + \frac{1}{2}mv_{cyclotron}^2$	No (standing wave solution)

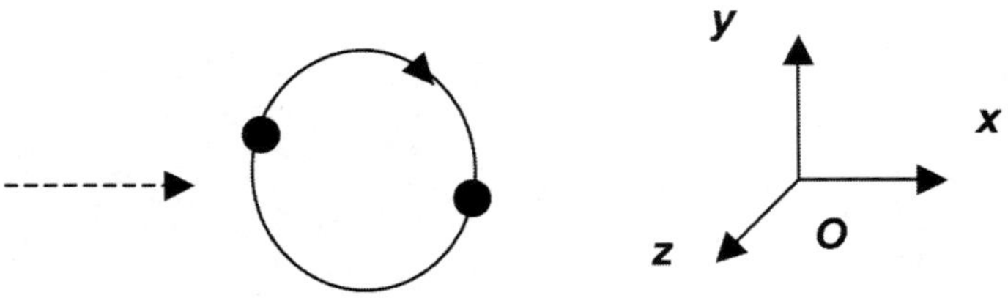

Figure A.9. Composite photon moving past point 0 along x-axis rotating in z–x plane.

proposed by Heisenberg in 1927. We present HUP in the original form given by Heisenberg. Assume the wave packet illustrated by Fig. A.10 consists of sinusoidal plane waves of wavelengths close to λ_0. There are approximately $n = \Delta x/\lambda_0$ wavelengths within the packet. Outside the limits of the wave packet the waves must cancel each other by interference. This only occurs if at least $n+1$ waves fall inside the limits of the wave (Fig. A.10). Thus we may write

$$\frac{\Delta x}{\lambda - \Delta\lambda} \geq n + 1 \tag{A.22}$$

In (A.11), $\Delta\lambda$ is the approximate range of wavelengths. Thus,

$$\frac{\Delta x \Delta\lambda}{\lambda_0^2} \geq 1 \tag{A.23}$$

The group velocity of the wave can be written as

$$v_g = \frac{h}{m_p \lambda_0} \tag{A.24}$$

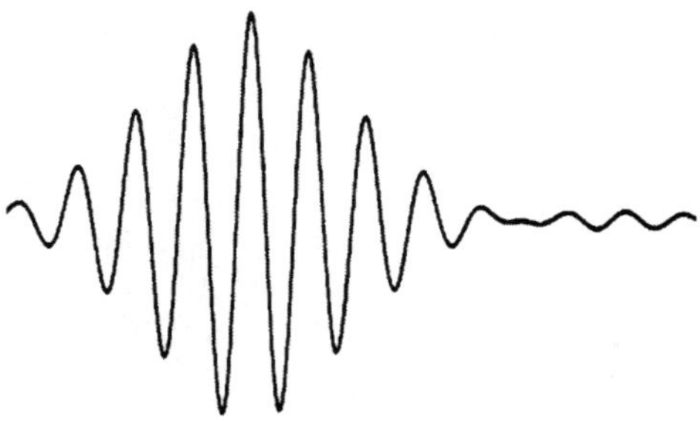

Figure A.10. Wave packet with uncertainty Δx in position and Δp in momentum.

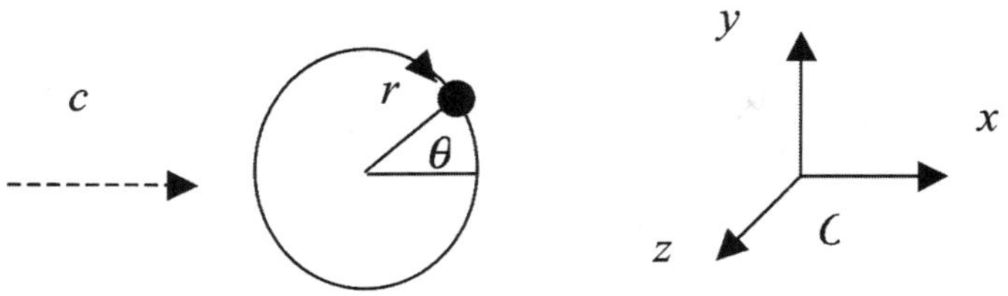

Figure A.11. Single particle photon moves past point 0 along x-axis rotating in x–y plane.

The spreading of the wave is thus characterized by a range of velocities

$$\Delta v_g = \frac{\hbar}{m_p \lambda_0^2} \Delta\lambda_0 \tag{A.25}$$

By definition $\Delta p = m_p \Delta v_{gx}$ and hence

$$\Delta p_x \Delta x \geq \hbar \tag{A.26}$$

HUP may also be written in the corresponding form

$$\Delta t \Delta E \geq \hbar \tag{A.27}$$

From Fig. A.11 we might model the wave packet in a somewhat unphysical fashion, as a single particle rotating around a centre of motion while translating along the x-axis. In this case uncertainty Δx in the x-direction involves a trigonometric function in θ and the scalar radius r of the photon's internal motion. Δx is seen to range from positive to negative in keeping with the wave motion. Similarly, the uncertainty in momentum Δp is also a function of θ and r as well as the particle's mass and also ranges from negative to positive. Trying to investigate uncertainty in this case leads to an unphysical result. We can either assume this is because there is something wrong with the concept of photon mass or perhaps the model is unrealistic. It is only when the photon is modelled by two particles as in Fig. A.10 the uncertainty function becomes balanced by the symmetry of the sub-photonic particles and uncertainty goes to nett zero. HUP is thus an incorrect model of the photon and can be considered a theoretical error leading to numerical inaccuracy that mandates the need for re-normalization. We see that the theme of SFT as a mutual effect between two particles obviates uncertainty, rendering the motion of the photon deterministic.

As an epilogue to this discussion the equations of HUP can be seen to be closely related to the central equations of SFT (2.24)–(2.25) where the motion of the electron is analyzed. These equations are exact whereas in HUP they are inequalities. In this case the motion of the electron is modelled via an infinite mass proton, similar to the one-particle photon model of Fig. A.11 but not unphysical in the sense that this is a reasonable approximation to the physics.

Note that Heisenberg assumes neither the group mass is a variable depending on relativistic speed nor is it zero. Heisenberg prefaces the derivation of his principle of uncertainty:

> The uncertainty relation specifies the limits within which the particle picture can be applied. Any use of the words "position" and "velocity" with an accuracy exceeding that given by equation (A.26) and (A.27) is just as meaningless as the use of words whose sense is not defined.

In a footnote he continues:

> In this connection one should particularly remember that the human language permits the construction of sentences which do not involve any consequences and which therefore have no content at all-in spite of the fact that these sentences produce some kind of picture in our imagination; e.g., the statement that besides our world there exists another world with which any connection is impossible in principle, does not lead to any experimental consequence, but does produce some kind of picture in the mind. Obviously such a statement can neither be proved nor disproved. One should especially careful using the words "reality," "actually" etc., sine these words very often lead to statements of the type just mentioned.

It needs to be stressed that while uncertainty can be zero this tells us more about the model's theoretical basis; it tells us little about how the model stands up to experiment. It is this observation that leads us to conclude that uncertainty is a form of theoretical inaccuracy as distinct to a numerical inaccuracy. It was Heisenberg's opinion that the photon could not be understood in a deterministic fashion. While he was correct that any attempt to use photons to examine a photon would disturb the experiment, this does not prevent us from examining the insides of the photon mathematically. Experimentally we may simply use numbers of photons to test individual photons. When we see blue light we are not perturbing all photons with blue wavelengths but only those photons that reach our eyes. Hence we can assume the vast majority of the photons we perceive will have an associated energy.

SFT and ZNR yield exact equations rather than the inequality equations (A.26)–(A.27). Heisenberg's statements seem more about an afterlife rather than scientific comments about uncertainty being a fabric of reality which was their context. If one were able to ask their great-great-grandmother whether she believed in the reality of television her answer would depend on whether she had ever lived in the age of television. Otherwise she would have denied such a preposterous suggestion, and her view would have been supported by thousands of years of human experience. Similarly, quantum mathematicians might deny the existence of deterministic equations of motion for the electron in the atom. Depending on whether they knew of the modern variant of SFT, they might think physics was a fixed and constant commodity, as immutable as life on Earth, or perhaps more pertinent Planck's "constant." Now we can perceive a complete world of theory inside the photon that promises to yield new physical understandings. Perhaps this HUP episode of science in the early and late 20th century gives us a more complete understanding of the differences between modelling and reality. We accept the assumptions of a scientific model on faith in order to find a way forwards via theory to an understanding of reality. The more two-way reflections there are between physics and its theory the more refined becomes our perception. We must keep our cerebral antennae open to outside broadcast without closing off reception.

One last comment can be made about the mathematics of uncertainty and the equations of SFT. Both are remarkably similar and it remains a historical quirk of fate that SFT did not emerge before WWII. It may be that the scientific world does not move with any haste. That may be been both a good and a bad thing – good in that no presumptuous errors are made but bad in that the world has to wait eighty odd years to finally begin to peer inside the photon with all the physical and biophysical structure that this knowledge contains.

A.8 Over-constraint of Classical and Quantum EM Equations

Throughout this book the central thesis has been the need for modelling the photon via a bi-spinorial function of motion. Given by (2.8) this relationship is written as

$$r\left(r_o, \omega_o, r_c, \omega_c\right) = r_o e^{j\omega_o t} + r_c e^{j\omega_c t}$$

Prior to SFT a mono-spinorial model rather than a bi-spinorial model was considered as in the Hertzian potentials of the cross dipoles examined earlier in this appendix. In this case, the dipoles are considered separate physical entities rather than two interacting atomic particles. We only come to an equation of dynamic balance via equating the two radiating fields and finding the associated spinors.

Looking at the problem as a system of over-constrained equations take the general case of a linear system of equations $A_n^m \vec{x}_n = \vec{b}_n$

(a) where the system is square of order $m = 4$ and $n = 4$

$$\begin{bmatrix} a_{11} & a_{12} & a_{13} & a_{14} \\ a_{21} & a_{22} & a_{23} & a_{24} \\ a_{31} & a_{32} & a_{33} & a_{34} \\ a_{41} & a_{42} & a_{43} & a_{44} \end{bmatrix} \begin{bmatrix} x_1 \\ x_2 \\ x_3 \\ x_4 \end{bmatrix} = \begin{bmatrix} b_1 \\ b_2 \\ b_3 \\ b_4 \end{bmatrix} \tag{A.28}$$

The system may have four real roots leading to physical solutions

(b) where $m < n$, e.g. $m = 3$ and $n = 4$

$$\begin{bmatrix} a_{11} & a_{12} & a_{13} \\ a_{21} & a_{22} & a_{23} \\ a_{31} & a_{32} & a_{33} \end{bmatrix} \begin{bmatrix} x_1 \\ x_2 \\ x_3 \\ x_4 \end{bmatrix} = \begin{bmatrix} b_1 \\ b_2 \\ b_3 \end{bmatrix}$$

The system is under constrained and there is an infinite number of solutions

(c) where $m > n$, e.g. $m = 4$ and $n = 3$

$$\begin{bmatrix} a_{11} & a_{12} & a_{13} & a_{14} \\ a_{21} & a_{22} & a_{23} & a_{24} \\ a_{31} & a_{32} & a_{33} & a_{34} \\ a_{41} & a_{42} & a_{43} & a_{44} \end{bmatrix} \begin{bmatrix} x_1 \\ x_2 \\ x_3 \end{bmatrix} = \begin{bmatrix} b_1 \\ b_2 \\ b_3 \\ b_4 \end{bmatrix}$$

The system is over-constrained and there are no solutions. However we can instead solve the over-constrained case by choosing a metric for instance least squares, the Pythagorean distance, to minimize the error. This example is a direct analogue to the case of quantum theory where uncertainty is present and the photon is modelled without taking its internal structure into account. In case (a) exact solutions exist as in the case of SFT. Note that the exact solution due to SFT may not be a true representation of the physical world due to some as yet unforeseen modelling errors. However, this is left for the future theoretical physicists who may well bump up against the next layer of the knowledge onion.

Appendix B

Comments on Physical Constants, Equations, and Standards

Table B.1 lists some of the important physical constants given by the U.S. standards body National Institute of Standards and Terminology (NIST). The basis of these values goes back to the earliest days of electric and magnetic theory and measurements conducted by Coulomb and Ampere that occurred prior to the theories of EM and relativity by Maxwell and Einstein. They reflect the underlying unity of electric and magnetic phenomena that was discovered by Maxwell around 1864. In some ways they reflect the world view of science after 1887 when Michelson and Morley performed their famous experiment to detect any ether throughout space. Einstein's postulate that the speed of light was constant only served to set these ideas of a fixed cosmos in neurological concrete. Many now believe that the constancy of the speed of light is an experimental measurement of the speed of light as observed from Earth. The speed of light varies depending on the constitutive parameters of the medium in which light travels such as the tissues of the body. This alters both the wavelength and the energy of an EM wave as it propagates through a general medium. Some say the speed of light is *defined* to be a constant. This is a matter of legality, often related to

Self-Field Theory: A New Mathematical Description of Physics
By Tony Fleming

www.panstanford.com

international agreements or commercial need. It is not a matter for physicists who wish to progress their physical understanding.

The permeability of free space μ_0, sometimes termed the magnetic constant, comes from a special case of Ampere's law where two wires of equal length carrying equal currents impose magnetic force on each other

$$F_{\mathrm{m}} = 2k_{\mathrm{A}}\frac{I_1 I_2}{r} \tag{B.1}$$

where $k_{\mathrm{A}} \equiv \mu_0/4\pi$ and $\mu_0 \equiv 4\pi \times 10^7\ \mathrm{N}/\mathrm{A}^2$ in SI units. This particular value for μ_0 is a result of the original pre-1900 units being centimetre-gram-second (CGS) and electromagnetic units (EMU) while the current standard is metre-kilogram-second (MKS) and the International System of Units (SI). Rather than being an intrinsic measurement conducted in free space, the value is a conversion from old to present standards in order to retain the Ampere as the unit of steady current. The permittivity of free space ε_0 is evaluated via the theoretical expression $\varepsilon_0 = (\mu_0 c^2)^{-1}$, where the speed of light is defined to be a constant as is μ_0. These definitions for c and μ_0 tend to make these values less concerned with experiment and more concerned with legal writ and that is a serious problem besetting standards at present.

According to SFT, in generalized media there is a range of values for h and c other than those of free space, and these are influenced by the constitutive parameters ε, and μ. This puts the variability given by $\varepsilon_0 = (\mu_0 c^2)^{-1}$ on a different footing. The values for $\hbar$ and c relate to phenomena in photonic, atomic, and nuclear physics and require careful measurement in standards and high energy laboratories. Again there is pressing need to clarify the current methods of measurements of complex permittivities and permeabilities in a range of applications including high energy regions and biological media. In biological media, magnetic interactions are diminished by setting permeabilities in tissues to that of free space, for example cells. In biological media lumped circuit analysis is sometimes used e.g. the Maxwell-Wagner model. In their day such models were state of the art, however they are now seen as intrinsically flawed. Tissues can radiate and hence a complete EM or photonic analysis is required to detect the so-called non-thermal effects.

In strong nuclear regions, SFT indicates EM and acoustic effects need to be considered together. As discussed in Chapter 4 there is currently a wide range of EM-acoustic effects that need to be integrated under the one theoretical umbrella. SFT suggests two extra homogeneous equations to link electric, magnetic, and acoustic effects together under the one set of modified ML equation in strong nuclear regions. If we choose to link the electric and the acoustic effects together and where the region is assumed isotropic and homogeneous the two equations are

$$\nabla \bullet \vec{N} = 0 \tag{B.2a}$$

$$\nabla \times \vec{N} + \nu_n \frac{\partial \vec{E}}{dt} = 0 \tag{B.2b}$$

where the modified Lorentz equation for the forces acting on the quarks is written

$$\vec{F} = q_q \vec{E} + q_q \vec{v} \times \vec{B} + q_q \vec{v} \times \vec{M} \tag{B.2c}$$

and the constitutive equations are $\vec{B} = \mu_n \vec{H}$, $\vec{D} = \varepsilon_n \vec{E}$, and $\vec{M} = \nu_n \vec{N}$, where ε_n, μ_n, and ν_n are invariant scalars, the nuclear constitutive parameters being similar to those of free space, ε_0 and μ_0, except the energy density within the nucleus depends upon the volume density of the gluon fields

$$dU_N = \rho_N dV = \frac{1}{2}(\varepsilon_n \vec{E} \bullet \vec{E} + \mu_n \vec{H} \bullet \vec{H} + \nu_n \vec{N} \bullet \vec{N})dV \tag{B.2d}$$

where $\vec{N}$ is termed the nuclear field and $\tilde{M}$ is the nuclear flux density. It is noted that another pair of homogenous equations could be written in terms of the magnetic and the acoustic effect

$$\nabla \bullet \vec{N} = 0 \tag{B.2e}$$

$$\nabla \times \vec{N} + \alpha_n \frac{\partial \vec{H}}{dt} = 0 \tag{B.2f}$$

In these two equations there is a variable α_n, which is a magnetic–acoustic parameter relating to the coupling between magnetic and nuclear fields. It seems prudent at this point in time to avoid the terminology and nomenclature of acoustic and vibrational effects in free space and to call the new third field a nuclear field even though it is theoretically likely that the nuclear field is a phonon field equivalent to the longitudinal component of the EM field. If such a system of modified ML equations can be detected by measurement inside the nucleus then the six equations given in Chapter 2 could be used for the strong nuclear effect.

In summary, there is a continuing struggle between standardization and progress. It appears that as soon as the community of scientists agrees on a standard someone else destroys the consensus by forging a new path into the future requiring a newer consensus. So the struggle must continue as scientific evolution emerges. In this regard knowledge often occurs in layers similar to an onion. Pertinently there are currently moves afoot to legally define the photon mass to be zero. As we have seen in this text there is every reason to suspect that such a move is based on the faulty concepts of the quantum theory that has been in service since 1927. SFT now demands a rethink about a range of fundamental questions concerning physics including uncertainty, the current QFTs *and* the mass of the photon.

Table B.1. Physical constants

Physical constant	Symbol	Expression	Known value[a]
Planck's number	h	$h = \frac{q^2}{2\varepsilon_0 v_e}$	$6.62606896 \times 10^{-34}$ Js
Speed of light	c	$c \equiv (\varepsilon_0 \mu_0)^{-1/2}$	$c \equiv 2.99792458 \times 10^8$ m/s
Fundamental unit of charge	q	$q = F N_A^{-1}$	$1.602176487 \times 10^{-19}$ C
Permittivity of free space (electric constant)	ε_0	$\varepsilon_0 = (\mu_0 c^2)^{-1}$	$10^7/(4\pi c^2)$
Permeability of free space (magnetic constant)	μ_0	$F_m = \mu_0 I^2/2\pi r$	$\mu_0 \equiv 4\pi \times 10^{-7}$ H/m
Rest mass of electron	$A_r(e)$ or m_e		$5.4857790943 \times 10^{-4}$ u $9.10938215 \times 10^{-31}$ kg
Rest mass of proton	$A_r(p)$ or m_p		1.00727647 u $1.672621637 \times 10^{-27}$ kg
Rest mass of photon	m_γ	$\frac{\hbar \omega_\gamma v_e}{4c^3}$	0.396×10^{-55} kg 0.221×10^{-19} eV
Atomic mass unit relative to C_{12}	u		1.660566×10^{-27} kg
Bohr radius	r_B	$r_B = \frac{\varepsilon_0 \hbar^2}{\pi m_e q^2}$	$0.52917720859 \times 10^{-10}$ m
Fine-structure constant	α	$\frac{v_e}{c} = \frac{4 m_\gamma c^2}{\hbar \omega_\gamma}$	$7.2973525376 \times 10^{-3}$ m

[a] CODATA values except photon mass extracted 19 July 2009 (http://physics.nist.gov/cuu/Constants/bibliography.html).

Appendix C

Self-Field Theory: New Photonic Insights[1]

A. H. J. Fleming
Biophotonics Research Institute, Melbourne, Australia
tfleming@unifiedphysics.com

Self-field theory is a new description of electromagnetic interactions. At its heart are bi-spinorial motions for both the electromagnetic fields and the interacting particles. Among its recent successes it has solved a simple model of the hydrogen atom, obtained an analytic estimate for the mass of the photon, and provided the first glimpses of structure within the photon. This may yield an organizational structure for bosons reminiscent in some ways to the chemical table that was glimpsed by Mendeleev in 1860 via a 2-D array of elemental properties. The self-field formulation obtains an analytic expression for Planck's number providing a basis for its understanding as a variable of motion applying equally to the electron, the proton, and the photon. While there are many differences, this report shows how the fields of self-field theory vary from classical electromagnetics and quantum field theory. In

[1]This Appendix is based on the original paper presented by A.H.J. Fleming, Self-Field Theory: New Photonic Insights, *PIERS-10*, Xi'an, Mar. 22–26, 2010, which was published and is available from the online conference proceedings at http://piers.org/piers2010Xian/

Self-Field Theory: A New Mathematical Description of Physics
By Tony Fleming

www.panstanford.com

classical electromagnetics the field covers all solid angles around a charge and is defined as a vector. Quantum field theory models the field as quanta shown as small wavy lines within Feynman diagrams; the mathematics does not specify an actual path, only the start and the finish points are specified where a Dirac-delta function is used to insert a propagator kernel or Green's function. Basically quantum field theory models the field as an impulse specified at space points. The uncertainty within quantum field theory is related to the lack of a complete electromagnetic bi-spinorial field form. The fields in self-field theory are discrete streams of photons, not like the continuous fields of Maxwell's classical electromagnetics. The photons are specified via a bi-spinorial function as spatially and time-varying motions including spiral-helices between the electron and proton of the hydrogen atom. Thus two distances are involved in the bi-spinorial motions, not one, fundamentally different to its forebears.

C.1 The Equations of CEM, QFT, and SFT

The history of classical electromagnetics (CEM) can be summarized by the equations of electro- and magnetostatics derived by Coulomb and Ampere (C.1a) and (C.1b) and their evolution over nearly a century to the Maxwell equations shown below as (C.3a)–(C.3d). The static equations describe uncoupled E- and B-fields while Maxwell's equations are coupled E-and H-fields and are first-order partial differential equations. Maxwell's original equations were 20 in number and the variables were potentials. Maxwell understood these equations as having an underlying wave basis and he succeeded in deriving wave equations showing the EM phenomenon was a plane-wave propagating in free space at the speed of light. The Linard–Wiechert potentials are taken directly between charge points and give the relativistic time-varying electromagnetic (EM) fields for point charges in arbitrary motion.

$$F_{\mathrm{E}} = \frac{q_1 q_1}{4\pi \varepsilon_0 r^2} \tag{C.1a}$$

$$\oint_C B \bullet dl = \mu_0 I_{\mathrm{enc}} \tag{C.1b}$$

The history of quantum field theory (QFT) began with the evolution of Schrödinger's non-relativistic equation to Dirac's relativistic equation. Schrödinger expressed the phase of a plane wave as a complex factor (C.2a). The Klein–Gordon equation (C.2b) is a relativistic version of Schrödinger's equation. Using Feynman notation (C.2c) we see how the Klein–Gordon equation can be analytically factorized to yield the positive and negative energy versions of the Dirac equation (C.2d) where the γ_μ are 4×4 bi-spinors. Dirac thus predicted the electron's anti-particle, the positron, via these equations. Although Dirac's equation is first order, it is derived from a second order wave equation and it is this that demands gauge symmetry and invariance under conformal transformations.

$$i\hbar\frac{\partial}{\partial t}(x,t) = \widehat{\mathrm{H}}\Psi(x,t) \tag{C.2a}$$

$$\left[\nabla^2 + \frac{m^2c^2}{\hbar^2}\right]\psi(r) = 0 \tag{C.2b}$$

$$(\partial^2 + m^2)\psi = 0 \rightarrow (i\partial + m)(i\partial - m)\psi = 0 \tag{C.2c}$$

$$(\gamma_\mu\partial_\mu + im)(\gamma_\mu\partial_\mu - im)\psi = 0 \tag{C.2d}$$

In self-field theory (SFT) the particles and EM fields that control the motions of charged particles satisfy the Maxwell–Lorentz (ML) equations. For application to atomic physics, regions where particle–field interactions occur are assumed isotropic and homogeneous and the constitutive parameters, ε_0 and μ_0 the permittivity and permeability of free space, are scalars. Where discrete particles carrying units of elementary charge q of opposite polarity are studied, in the absence of nebular regions of charge and current density, the ML equations are written as

$$\nabla \bullet \vec{E} = \frac{q}{v_q} \tag{C.3a}$$

$$\nabla \bullet \vec{H} = 0 \tag{C.3b}$$

$$\nabla \times \vec{E} + \mu_0\frac{\partial \vec{H}}{dt} = 0 \tag{C.3c}$$

$$\nabla \times \vec{H} - \varepsilon_0\frac{\partial \vec{E}}{dt} = \frac{\pi}{s_q}q\vec{v} \tag{C.3d}$$

The Lorentz equation for the field forces acting on the particles is written as

$$\vec{F} = q\vec{E} + q\vec{v} \times \vec{B} \tag{C.3e}$$

The constitutive equations in free space are

$$\vec{B} = \mu_0 \vec{H} \tag{C.3f}$$

$$\vec{D} = \varepsilon_0 \vec{E} \tag{C.3g}$$

The relationship between the speed of light and the ratio of the fields

$$c = (\varepsilon_0 \mu_0)^{-1/2} \tag{C.3h}$$

$$dU = \rho dV = \frac{1}{2}(\varepsilon_0 \tilde{E} \bullet \tilde{E} + \mu_0 \tilde{H} \bullet \tilde{H})dV \tag{C.3i}$$

The atomic energy density per volume (C.3i) depends upon the E- and H-fields in the atomic region. Equations (C.3a)–(C.3d) are termed the EM field equations. In these equations, $v$is the particle velocity, m is its mass. The motion of the electron forms an EM self-field solution, its position is a sum of two spinors $\sigma_o(r_o, \omega_o)$ and $\sigma_c(r_c, \omega_c)$, where r_0 and r_c are orbital and cyclotron radii and ω_0 and ω_c are orbital and cyclotron angular velocities.

$$r\,(r_o, \omega_o, r_c, \omega_c) = r_o e^{j\omega_o t} + r_c e^{j\omega_c t} \tag{C.4}$$

In equation (C.4) only discrete eigenvalues of r_0, r_c, ω_0, and ω_0 are allowed. After suitable algebraic manipulation the system may be recast in terms of the energies in matrix form. The principal mode can be written as

$$\begin{bmatrix} 1 & 0 \\ 0 & 1 \end{bmatrix} \begin{bmatrix} V \\ T \end{bmatrix} = \begin{bmatrix} \hbar\omega \\ \hbar\omega \end{bmatrix} \tag{C.5}$$

Here $V = V_o = \frac{1}{4\pi\varepsilon_0}\frac{q}{r_o} = V_c = \frac{1}{4\pi\varepsilon_0}\frac{q}{r_c}$, and $T = T_o$, $T_c = \frac{1}{2}m_e v_o^2 = T_c = \frac{1}{2}m_e v_c^2$ are orbital and cyclotron components of the potential and kinetic energies. In the principal mode $\omega = |\omega_o| = |\omega_c|$ and $r = |r_o| = |r_c|$, where the orbital and cyclotron velocities are also equal $v_o = \omega_o r_o = v_c = \omega_c r_c$. The right-hand side of (C.5) contains a variable termed Planck's number that empirically agrees with the known value of Planck's constant to an accuracy of seven significant figures. The solution of (C.5) agrees with similar accuracy to the

Bohr radius and the known resonant frequency of the hydrogen atom. By involving the effect of the magnetic field upon the electron this solution forms an extension of Bohr's theory. As it stands (C.5) contains four variables yielding the deterministic motion of the electron given by (C.4). This agrees with the four quantum numbers known via quantum mechanics.

If the intrinsic energy of the system changes from that of free space ε_0 and μ_0 (C.3f), (C.3g), and (C.3i) the EM fields within the atom can adjust altering the atomic and molecular binding structures. In effect this means that the radial and spin states of the two photons involved in the binding energy can adapt to the energy change where the photons themselves are assumed to have a composite structure. Now the system has six degrees of freedom including the E- and H-fields in response to the change in ambient energy. The two extra variables give a range of variation orthogonal to the phase diagrams of atoms. A typical phase diagram shows only a solid line separating the various phases. SFT indicates a small range of binding structures that depend on the intrinsic energy of the system.

C.2 The Electric and Magnetic Fields of CEM, QFT, and SFT

The photon streams of SFT are taken between centres of rotation, not between charges as in CEM and QFT. Thus two distances are involved in the bi-spinorial motion and not one. The fields in SFT are discrete quanta, photons, rather than the continuous vector fields of CEM. QFT also models the fields as discrete quanta. However each individual field in QFT may be taken over the entire range of solid angles connecting any number of particles depending on the geometry and the number of interacting atomic particles. The field can be shown as a small wavy line within Feynman diagrams but the QFT mathematics does not specify any actual path, only the start and end points where Dirac-delta functions are used to insert propagator kernels and Green's functions depending on the field and its geometry. In QFT the field is actually a potential, not the E- and H-fields of SFT. In QFT the fields (potentials) are modelled as impulse functions specified at charge points. Within SFT, a pair of particles

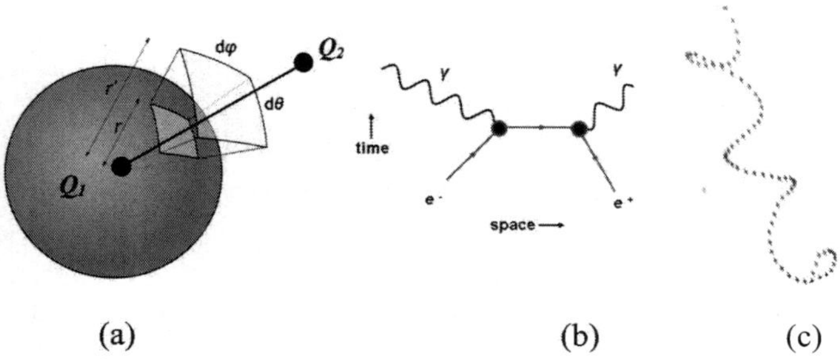

Figure C.1. Field forms (a) CEM, (b) QFT, (c) SFT. See also Color Insert.

defines the bi-spinorial field and its motion, this pair of particles and its bi-spinorial field form a unique couple. Due to orthogonality the fields of this couple do not influence any other charges apart from the couple. The transit of the SFT field is specified via the bi-spinorial function and assumes various motions including spiral-helices as it transits between the electron and proton within the hydrogen atom. Certainly there is a vast difference between this SFT time-variant field motion and the time-invariant CEM where the field ubiquitously covers all solid angles with no definition other than its vector nature as to the actual field motion, field flux being the only indicator of field motion. Similarly the uncertainty of the field within QFT is related to its lack of a complete and coupled EM bi-spinorial field form.

There are other major differences including an absence of Heisenberg's uncertainty principle (HUP) within SFT. As the photon is modelled via bi-spinor uncertainty is obviated. In SFT the electron's self-fields are modelled via a complete EM function that explicitly includes both E- and H-fields, enabling the complete analysis of the mutual self-field effect between two particles. Unlike the quantum potentials that are expectations yielding probabilistic solutions, the bi-spinorial field variables of SFT allow completely deterministic solutions. This results in a clearer picture of the physics that includes the particleα–photon interactions and the binding mechanism. The solution is complete (coupled) and based on the first-order ML equations, hence neither special relativity nor gauge symmetry is problematic. The bi-spinor field variables of SFT are a priori relativistic and its solutions analytic rather than numerical. In EM applications gauge symmetry is maintained by analyzing pairs of particles and not just single particles. Similarly, a pair of conjugate photons of finite mass do not constitute *symmetry breaking* of the Lagrangian as in QFT.

References

1. Jackson J. D., *Classical electrodynamics*, 3rd Edition, John Wiley & Sons, New York, NY, 1999.
2. Miller, A. I. *Early Quantum Electrodynamics, A Source Book,* Cambridge University Press, Cambridge, U.K., 1995.
3. Fleming, A. H. J. Electromagnetic self-field theory and its application to the hydrogen atom, *Physics Essays*, Vol. 18, No. 3, 265–285, 2005.
4. Fleming, A. H. J. Self-field theory, analytic spectroscopy of the ordinary photon, *Proc. 2nd EHE Intl Conf.*, Wroclaw, Poland, 18–23, 2007.
5. Fleming, A. H. J. Analytic Estimate of the Mass of the Photon, *PIERS-09*, Moscow, Aug. 18–21, 2009.

Appendix D

Frequently Asked Questions

Self-field theory (SFT) entered the scientific lexicon and mainstream of peer review a few years ago just after the turn of the new millennium. Although it is still going through this process of validation, SFT has a less contrived more natural heritage as a mathematical physics than its much older companion theory quantum field theory (QFT). This appendix puts together a list of questions that have been posed either during presentations, over the internet on physics forums, or similar physics discussion groups. It is clear to the author that those who are more physicist than mathematician may not recognize SFT as the breakthrough in mathematical physics that it indeed represents. Many theoretical physicists are trained almost exclusively on concepts of Lagrangians, gauge theory, symmetry, and a grab bag of specialized subjects associated with the current versions of quantum theory that are based on second-order wave equations and the integral equation techniques used to solve them probabilistically.

While there is nothing whatsoever wrong with integral equation methods, what is at issue is an over-constrained model of the photon as a point source. This introduces a numerical error known as uncertainty and the present day probabilistic nature of quantum theory. In addition the higher the order of the mathematics the

Self-Field Theory: A New Mathematical Description of Physics
By Tony Fleming

more complex it becomes. This applies especially to differential forms of equations. So the question boils down to whether to use the electromagnetic (EM) wave equations or the first-order Maxwell—Lorentz (ML) equations as the basis of the mathematics. For instance many queries concern either gauge or symmetry. There is no such thing as a gauge theory in SFT because of the simplistic way gauge applies within SFT; further there are no vector and scalar potentials in SFT only the E- and H-fields. Again, symmetry is achieved in a different way in SFT compared with QFT. Hence the issue of a non-zero-mass photon is also answered in a mathematically different fashion to the symmetry-breaking presence of non-zero-mass fields within QFT. Finally relativity is inherent in the bi-spinorial form of mathematics and provides a novel way to incorporate relativity within quantum theory. This also changes the way we view relativity to be physically palatable. Relativity's warping of space-time is intuitively digestible when it includes the hidden time and distance of internal photon motions.

There is an enormous amount of validation yet to be performed relating to SFT. This work needs to be "delegated" to other than those who are aware of the predictive power of SFT similar to the case in the late 1920s and early 1930s with quantum mechanics (QM) and the early concepts of quantum electrodynamics (QED) when pioneers in various disciplines emerged to find new predictions and validate QFT in the various areas of physics to which it could successfully be applied.

Within this book the statement "according to SFT" is frequently used. This usually implies a hypothesis that is yet to be validated by experimental evidence. However, this does invariably indicate a physical process that can be observed to correspond to the predicted mechanisms of the theory qualitatively without comprehensive quantitative validation. Photonics within the atmosphere is one example. What we see are layers of chemical reactions, each different to the others. Anyone in an aircraft either in the process of attaining cruising altitude or descending in preparation for landing will have observed such layers of photonic interaction. It is the author's role to simply present the work thus far including both the evidence garnered to date and the many hypotheses or predictions that can be associated with SFT across the widest possible gamut of physics and biophysics. That is what a theory of mathematical physics is – a means of predicting the physical world. The process of science is to improve on our theories. At this point in time it is our quantum theories first mooted around the start of the 20th century that are being questioned and found wanting in their current form. In the fullness of time, SFT will face the same curtain call. But we probably have some time yet before we come to that bridge.

The author is an applied mathematician with an engineering and biological background, a lifelong polymath who has discovered an exact solution to the ML system of equations that has previously remained concealed. This is the heart of SFT, an important new analytic theory of how various physical systems of particles can remain in dynamic equilibrium without expending energy. SFT turns out to be very general, not just applying to EM particles as in the hydrogen atom. SFT provides a new description across the known limits of physical knowledge. There is overwhelming evidence that SFT is a more detailed more fundamental model of physics than QFT. In the past we have spoken of a unification of physics and that is precisely what SFT represents. It brings all four known forces under the one umbrella, EM, strong and weak nuclear, and gravitational. It predicts new gravitational forms at the galactic and higher levels of cosmological structure.

The author has travelled widely around the globe over the past decade in an effort to spread knowledge of SFT as widely as possible. Occasionally he is asked about the evidence for SFT's predictions. While the evidence remains qualitative in the main, there is some astounding evidence that will surprise theoretical physicists who have yet to learn of SFT. To experts in a particular field, born and bred on the many intricacies of QFT, it may appear somewhat presumptuous as the many predictive assertions of SFT are outlined. That is what it should be. The scientific method is sceptical of new theories. Nothing is assumed without both a theoretical mechanism and a quantitative validation. Even the author's colleagues of bioelectromagnetics are nothing if not sceptics when it comes to new bio-effect mechanisms. For decades several bioelectromagnetic phenomena have been reviewed as unsubstantiated because there has been no theoretical mechanism — "good experimental evidence but we do not understand how such a bio-effect happens" — a form of Catch 22 that Capt. Yossarian knew all too well. Now we have such a mathematical mechanism that may be behind many possible biological phenomena and we need to complete the work of validation by matching theory with experiment. This is all work to be done within but one very important area of expertise. Speaking metaphorically, we are at the entrance of a mathematical tunnel leading to a new scientific and physical world, a "wormhole" leading to the future. It is like the layers of a knowledge "onion"; the wormhole leads us to a new level of existence.

It is hoped that the presentation in this textbook of the mathematics of SFT is convincing in its own right. The author has always sought to understand the mathematics at a physical level. Indeed SFT provides a much more physically intuitive picture of EM interactions than either classical electromagnetics (CEM) or quantum theory. It is hoped that this list of questions and answers will be useful to those who venture into SFT without the necessary mathematical depth and background. One good thing in favour of SFT is that the mathematical level required to comprehend it is several orders of complexity simpler than QFT and hence can be understood by undergraduates and the casual reader with at least two years of undergraduate mathematics. SFT may be unpalatable to the advanced exponent of QFT who has invested years of hard mental effort to master all the hand-waving, and chicanery necessary to learn all the nuances of QFT. In comparison, SFT is like a stroll in the park with an old friend. This should not be taken to mean SFT is classical. SFT seems to indicate that all physics and biological life involves an infinite series of energy levels or dynamical systems of particles. Thus physics and biophysics are fractal where each level is connected to a higher level, a lower level and within levels. Streams of field particles act as connectors between levels. This implies that there is an initial stream going back to the Big Bang and the absolute point in space-time which relates to all motion and life thereafter. SFT is a clearer mathematical model of reality than QFT.

What Is SFT?

SFT is a theory of mathematical physics that appears to apply across many, if not all, known areas of physics and biophysics. It is an extension to the present theoretical physics that achieves unification across current knowledge of physics and biophysics. At its most fundamental it is a series of analytic solutions to a range of physical applications across EM, weak and strong nuclear, and gravitational physics.

What experimental evidence exists for such claims?

(a) The primary evidence comes from the eigensolutions to the hydrogen atom that includes a theoretical derivation of Planck's constant $\hbar$. It should be noted that QFT is a heuristic formulation that proposed an eigenvalue system of equations based on the experimentally obtained estimate of $\hbar$ the eigensolution first determined by Planck in 1900. Hence SFT provides the theoretical eigenvalue basis for QFT.

$$Ax = \lambda x \tag{D.1}$$

$$\widehat{\mathrm{H}}\Psi(x,t) = i\hbar\frac{\partial}{\partial t}(x,t) \tag{D.2}$$

$$m_e v_o^2 = 2\hbar\omega_o \tag{D.3}$$

The three equations (D.1)–(D.3) demonstrate the situation described above. Equation (D.1) is the usual form of a general mathematical eigenvalue equation. Equation (D.2) is the Shrödinger equation where the left- and right-hand sides have been transposed left to right and right to left, to demonstrate the fact that Planck's (reduced) constant $\hbar$ is inserted heuristically from experiment in QM. $\hbar$ (Planck's reduced number) is a consequence of the formulation in (D.3) and does *not* come from experiment. If we re-work SFT in the wave equations of QFT, this would result in a new formulation for QFT one that was deterministic, not probabilistic.

(b) The equations of SFT are identical with those of uncertainty (HUP) except the inequality condition is replaced by *two* equality conditions one relating to the motions due to the electric currents the other relating to the motions due to the magnetic currents of each particle.

$$\Delta p_x \Delta x \geq \hbar \tag{D.4}$$

$$m_e v_o^2 = 2\hbar\omega_o \tag{D.5}$$

$$m_e v_c^2 = 2\hbar\omega_c \tag{D.6}$$

It turns out that in EM applications only two particles can be connected to each other. They are connected by a stream of photons. Each domain from cosmological and above, down to photon and below can only connect to a small discrete number of particles (two, three, or higher).

(c) While Bohr's theory of the electron is still recognized as a useful theoretical model of the electron in the hydrogen atom, it does not include a theory of the electron's magnetic field, nor the proton's. SFT supplies the missing theory for this simple model. In terms of SFT Bohr theory is said to be "mono-spinorial," while SFT is bi-spinorial. SFT "completes" the Bohr model of the electron and the proton. The analytic solutions obtained from SFT apply to all modes of the hydrogen atom.

(d) The bi-spinorial basis of SFT can be seen to be a form of relativity where a "hidden variable," the photon's internal motion, accounts for some of the motion treated as external distance and time in relativity. This internal distance and time causes the dilations observed by Einstein.

Where does SFT derive from?

SFT is an exact or analytic solution that comes from the existing theory of partial differential equations. It is important to note that this analytic solution can be applied to all states of the hydrogen atom, and is capable of being applied to *all* atomic and molecular structures. This may seem ludicrous or amazing to present-day chemists and scientists who use numerical methods to solve molecular structures. However, there is no mathematical reason why such a situation cannot be true. In fact SFT is capable of obtaining analytic solutions to nuclear physics. Again this is hypothesis and the work is yet to be performed. What can be said is that the author has obtained an analytic solution to the hydrogen diatom, commonly known as hydrogen gas, presented briefly in Chapter 3.

Is SFT related to QFT?

Yes, intimately, even though it has a much shorter history. It is based on the equations of CEM that could not be successfully applied to the hydrogen atom at the turn of the 20th century. It has taken another century to discover what was wrong with CEM and to solve the hydrogen atom analytically. Briefly, SFT is based on the first-order ML system of equations, while QFT is based on the derived second-order system of equations. Current QFT uses a different model of the photon to SFT. The photon in QFT is mono-spinorial while it is bi-spinorial in SFT.

Why is SFT not a Lagrangian formulation?

Because the Maxwellian basis is much easier to work with in the same way that finite difference method of numerical solution is far easier to implement than the finite element method. It is more closely linked at an intuitive level to the important role the photon plays in physics and biophysics, and to the geometric concepts seen in nature of rotation, helix, spiral, etc.

What role do gauge and symmetry play in SFT?

Gauge theory is almost trivial in SFT and as such there is little need to consider gauge apart from recognizing that each spinor rotation is $O(2)$ and non-unitary. When we consider two particles we need to consider two field particles. In the atom this boils down to the electron, the proton, and *two* photons. When we consider three particles we need to consider three field particles. In the nucleus this is three quarks (e.g. two up: up, down; one down: down, up), and three gluons.

What does SFT mean for EM? (What is wrong with CEM?)

The usual method of obtaining relativistic fields is via the Liénard Wiechert scalar and vector potentials. In other words, the fields are determined by taking the distance between two charges. In SFT this single distance is replaced by *two* distances taken between the centres of the bi-spinorial rotations of the photon. In other words there is not a single distance but two distances taken in measuring the fields. This is discussed in Chapters 1 and 3.

What does SFT mean for quantum theory in its various forms?

We need to differentiate between current implementations of QM and post-SFT implementations. If we change the way the photon is modelled inside quantum theory it will become a deterministic formulation. It may well be able to solve certain problems more accurately than SFT. Like their associated numerical methods, SFT and the modified quantum methods, there will situations where an integral equation method will be better suited to cover homogeneous regions and other applications where an inhomogeneous region will be better suited to SFT. It makes sense to retain the current nomenclature, i.e. quantum theory, rather than some new term with which workers in the field have to become accustomed. Any modifications, including different shape function for the photon, implicit incorporation of relativity via the bi-spinors, selection of mutual particles not single particles, and deterministic output should be seen overall as an organic process within the current terminology. That said quantum theory will change comprehensively though not totally. It may in time be seen to be worthwhile to rename quantum mathematical methods of analysis. Perhaps integral equation methods of SFT might be called integral SFT and the current SFT that acts on the Maxwellian and its modified forms might be called differential SFT. Whether the scalar and vector potentials or the E- and H-fields should be used as variables needs to be considered carefully to match a particular application. The largest change will be the use of analytic solutions in comparison with the present-day numerical solutions.

What does SFT mean for string theory?

Although SFT has other forms of forces it features a differential form of electromagnetics as a basis for gravity. The number of variables (dimensions) of space-time can be expressed as the three spatial variables and time or FOUR. We can term these by the following indices x, y, z, and t. Then we have the differentials. If we assume that time and distance never couple then the following differentials also apply xx, yy, zz, xy, xz, yz, and tt. This makes ELEVEN dimensions in total. This agrees with string theory. Hence there is an intimate relationship between SFT and string theory. This gives another rationale for the validity of SFT. The current problem with string theory is its lack of application to "everyday" or terrestrial physics. Since the basis of SFT is the point-particle we might easily change any point-particle to become a string or toroid of mass m and geometry (a toroid can be specified via two radii, r_1 and r_2) thus transforming SFT into the first string theory to cover the gamut of physics as with the point-particle version of SFT. This modelling manoeuvre overcomes the main problem with string theory, its lack of application to "everyday" physics. Whether or not matter behaves *totally* as strings is a matter for determination by modelling various situations including the hydrogen atom. The most appropriate answer may be a combination of strings *and* point-particles.

String theory can be seen to have originated from two underlying motives. First, it provides a way to achieve the wave-particle duality and secondly it overcomes the singularity at charge points within quantum theory, a numerical error; replacing a point with a small section of string achieves the required numerical fix within quantum theory. The problem with string theory is that it has been anchored to quantum theory with its CEM basis, namely its over-constrained model of the photon, and also therefore its uncertainty principle. Hence, like quantum theory string theory can yield the standard model of particle physics and is seen to be the way energy behaves at high energy levels. So, although it is an attractive method for mathematicians within particle physics to get their teeth into, it has to date remained somewhat removed from the everyday world of physics and remains within the ambit of quantum theory. It is not the author's desire to put hundreds and thousands of his colleagues within mathematical physics out of work. On the contrary, SFT can be seen to encompass both points *and* strings. The way photons act in their role as binding energy within atoms, molecules, all the way up to large complexes of atom as gravitational systems are fundamentally like "beads on a string." The same applies to gluons. Hence if we make the changes suggested above to quantum methods, the same changes in general apply to current methods of string theory. Then having made these modifications, string theory becomes a deterministic method involving closed-form solutions across physics. Most importantly string theory might then be applied within SFT or the modified quantum theory suggested above. At present only simple paths can be analysed within the current early forms of SFT, but we can foresee new methods of analysis encompassing both string theory and SFT, or string theory and the modified quantum theory. In this way string theory can be used within the wide gamut of physics to analyse the photon in its myriad roles as binding energy within atomic arrays. String theory thus becomes indispensable within future analytic methods to predict how atoms, molecules, and gravitational systems behave down to the photonic level.

At another level, if string theory can be used to replace photons, particles of light, then strings can be used to approximate a series of space-time points defining the motions of strings of photons or phonons. Thus they can be useful to simplify the mathematics, possibly within the nucleus. One essential difference between SFT and string theory is that SFT is fractal; it sees fields beneath the photon structure. Note that we have no reason to believe that the photon is *not* a particle, or an infinite series of particles that look like a single particle, and is instead a string. In fact we have experimental evidence beginning with Einstein's photoelectric effect that it is particulate. Nevertheless string theory may be useful in simplifying the fractal mathematics by concluding the infinite series in pieces of string, tying up loose ends so to speak. These differences become modelling or numerical issues.

What does SFT mean for cosmology and general relativity?

We have already seen that the internal and external motions of the photon help explain in a physically intuitive way how space is not actually warped but the vision our eyes see at relativistic speeds is warped. As Einstein knew, seeing is not always believing; straight lines could be curvilinear. In regards the general theory of relativity, Einstein's general relativity (GR) assumes a single form of gravitation acting across the entire Universe. To an approximation this is true but the actual situation is otherwise. SFT implies three main modifications to cosmological models based on GR as it currently stands: (1) Like CEM and its failure early in the 20th century to solve the atom, and quantum theory's lack of magnetic currents there is a lack of any stable solution due to the failure of models to examine mutual effects between masses. (2) The Universe contains more than one type of gravitation; a tri-spinorial form applies to galaxies, a tetra-spinorial form may apply to super clusters, and maybe another form perhaps a penta-spinor applies to the Universe itself. The overall structure of the Universe is therefore not homogeneous or isotropic as assumed by GR. (3) Another important modification relates to the photon's non-zero-mass and composite structure as described in Chapter 3. If the Big Bang was hot enough there would have been an initial period where a sea of sub-photonic particles existed. This may be responsible for the inflationary period when the Universe expanded to near its present size at superluminal speeds. Sub-photonic particles of non-zero mass could travel at superluminal speeds and help solve the so-called horizon problem in a more intuitive way. Similarly the anisotropy observed within the Universe can be explained without recourse to quantum foam theories. These modifications all have implications for the various GR solutions obtained by Friedmann, Lemaître, De Sitter, Guth, and others including Einstein's own solution obtained in 1915. Instead of the fluid dynamics of current GR models a particle-field model can give another perspective on cosmological processes. Overall this suggests an early inflationary period that finished before an evolution towards the critical condition on density leading to a dynamic equilibrium within the Universe. All effects are supported by the available cosmological evidence.

Historically, Einstein and Heisenberg crossed swords about GR and quantum theory.

A Conversation between Einstein and Heisenberg

Heisenberg: *"One cannot observe the electron orbits inside the atom. [...] but since it is reasonable to consider only those quantities in a theory that can be measured, it seemed natural to me to introduce them only as entities, as representatives of electron orbits, so to speak."*

Einstein: *"But you don't seriously believe that only observable quantities should be considered in a physical theory?"*

"I thought this was the very idea that your Relativity Theory is based on?" Heisenberg asked in surprise.

"Perhaps I used this kind of reasoning," replied Einstein, *"but it is nonsense nevertheless. [...] In reality the opposite is true: only the theory decides what can be observed."*
(translated from "Der Teil und das Ganze" by W. Heisenberg)
http://www.thebigview.com/spacetime/uncertainty.html

There are theoretical similarities between GR and quantum theory. Among the most fundamental are that both are based on second-order wave equations and their associated potential theories and gauge considerations. In comparison, SFT is based on the first-order Maxwellian with its field variables that have a much reduced emphasis on gauge. Both GR and QFT are based around single particle analyses rather than the mutual effects that couple particles together studied in SFT. Finally both GR and quantum theory employ a metric in the view of SFT to accommodate the over-constraint of the basic equations. In both cases this is linked to a theoretical requirement for a zero-mass photon.

Thus both quantum theory and GR depend upon a zero-mass photon and hence from the point of view of SFT both quantum theory and GR are theoretical approximations. For quantum theory zero-mass springs from the earliest observations of beta decay and again when a negligible rest mass of the photon could hardly be compared with the seemingly endless radiation from within the nucleus of the bombs dropped on Hiroshima and Nagasaki in 1945. The cosmological principle that had its genesis in the Vatican's unscientific and dogmatic dealings with Galileo was a way to avoid having any universal centre of gravity thus making the same mistake again. Nevertheless it is only an approximation in the light of SFT where it is seen that non-homogeneity and anisotropy are both present in the gravitational structure itself where space is divided into different gravitational regions. This structure depends on the composite nature and non-zero mass of the photon. The space within the Universe cannot be thought of as the surface of an expanding balloon other than as a theoretical approximation that holds for GR. It is known that at smaller than cosmological domains the cosmological principle does not hold for instance for any possible surviving location of the Big Bang. We may think of a biological tissue such as liver where the dielectric constant is averaged over the microstructure such as biological cells. While such an approximation is useful for numerical estimation it cannot be assumed to hold in any fine detail across smaller domains; this holds for both a homogenous isotropic model of liver and of the cosmos. This is not to denigrate GR in any way but to reveal the approximate nature of GR theory that Einstein had proposed in 1915 when it was and remains today a monumental intellectual achievement of the highest level within mathematical physics.

Appendix E

The Search for a General Physical Mathematics

Prior to SFT, physics, chemistry, EM, and many other areas of science depended on either quantum theory with its inherent limitations due to uncertainty or CEM with its limitation within macroscopic dimensions. Since before 1927 when quantum theory was introduced there has been an ongoing search for a general deterministic mathematics applicable across macroscopic and microscopic domains as well as the atomic and nuclear domains where probabilistic quantum methods have been preferred methods of choice. As mentioned in Chapter 1 Einstein rejected the probabilistic basis of quantum theory and searched unsuccessfully for a deterministic and unifying field theory that might be applicable across physics.

Another leading figure in this search was Herbert Fröhlich, who along with Albert Einstein can be considered the scientific forebears of all who have sought the path of a unifying mathematics. Fröhlich was involved in many areas of science including particle physics with contributions to nuclear forces and meson theory, and biology including dielectric theory and coherence. The author is grateful to Dr. Peter Rowlands of the Oliver Lodge Laboratory, Department of Physics at the University of Liverpool for a copy of the autobiographical book *Herbert Fröhlich FRS – A Physicist*

Self-Field Theory: A New Mathematical Description of Physics
By Tony Fleming

www.panstanford.com

Ahead Of His Time. Fröhlich was Chair of Theoretical Physics at the University of Liverpool from 1948 till his retirement in 1973 (from the position of Chair but not from further scientific endeavours until his death in 1991 at age 85). Fröhlich had used quantum theory in the area of semiconductors to great effect writing his first book *Elektronentheorie der Metalle,* Springer, in Berlin, 1936. Like many other German and Eastern European scientists of this era he fled persecution. He escaped St. Petersburg (then Leningrad) and the Stalinist purges of 1935 being interned as an "alien" in London in 1940. Many of his ideas including biological coherence have been taken up by others including the author whose work on diffusion within the membrane had its genesis in Fröhlich's ideas encountered during his Ph.D. Of particular interest to SFT is Fröhlich's previously unpublished "bilocal" extension of the Dirac Theory to take into account the magnetic currents in quantum theory. This is equivalent to the complete Bohr Theory solution of SFT that also takes into account the magnetic currents and motions of the electron. Thus Fröhlich and Hertz whose potentials in application to the half-wave dipole antenna were the original stimulus for the bi-spinorial basis within SFT can be closely associated with the origins of SFT.

On a recent trip to St Petersburg, Russia, to attend the International Congress-2010 on Fundamental Problems of Natural Sciences and Engineering, July, 26-30, 2010, the author met Vadim Matveev and his son Oleg who have discovered a macroscopic version of HUP indicating the existence of a macroscopic version of quantum theory. This work establishes the veracity of Fröhlich's suggestion some decades earlier. Their work is given here. The only difference between this work and SFT is the inclusion of the magnetic currents in SFT as mentioned above. In essence the existence of a macroscopic quantum theory demonstrates the wide applicability of both quantum theory and SFT.

UNCERTAINTY RELATIONS AS A CONSEQUENCE OF THE LORENTZ TRANSFORMATIONS

V. N. Matveev and O. V. Matvejev

Joint-Stock Company "Sinerta"
Savanoriu pr., 159, Vilnius, LT-03150, Lithuania
E-mail: matwad@mail.ru

Abstract

A macroscopic object consisting of a rod equipped with a pair of synchronized clocks is examined. General physical relations are directly derived from Lorentz transformations for the case of the rod's one-dimensional motion (along the X axis)—the uncertainty relation of the object's x-coordinate and the projection of its impulse along the X axis, p_x, and the uncertainty relation of the object's observation time, t, and its energy, E. The relations take the form $\Delta p_x \Delta x \geq H$ and $\Delta E \Delta t \geq H$. The H value in the relation has action dimensions and is dependent upon the precision of the rod's clocks and its mass. It is shown that if the macroscopic object itself performs the function of an ideal physical clock, the relations derived in the limiting case then take the form of $\Delta p_x \Delta x \geq h$ and $\Delta E \Delta t \geq h$, where h is the Planck's constant.

Introduction

The uncertainty relation of a micro-particle coordinate and the projection of its impulse along the coordinate axis, as well as the uncertainty relation of micro-particle time and energy, falls into the ranks of the most important quantum relations that express the known uncertainty principle in a mathematical form. Quantum relations are not observed in the macrocosm in the sense that they are vanishingly small with respect to macro-bodies and do not show themselves in practice. According to our data, no general physical relations have been reported in extant literature that extend to the macrocosm and that limit the simultaneous specification of the impulse and the coordinate of physical bodies. In light of existing notions concerning the properties of macroscopic bodies, the absence of such general physical uncertainty relations seems natural, since uncertainty relations have not been connected to measuring instrument errors, while it is possible to talk about macro-body coordinate, impulse, and energy inaccuracies, as they seem just as obvious today as measurement errors.

The objective of the work at hand was to demonstrate the existence of general physical uncertainty relations that extend to macro-bodies. This objective was achieved over the course of solving a problem that consisted of determining the velocity and the velocity-related physical values of objects based on the degree of desynchronization of moving clocks.

The relations $\Delta p_x \Delta x \geq H$ and $\Delta E \Delta t \geq H$ were directly derived from Lorentz transformations. The first relation connects the uncertainty of the projection of the impulse, p_x, of the body under consideration, Δp_x, to the uncertainty of the x-coordinate, Δx, while the second relation connects the uncertainty of the energy, E, of the object, ΔE, to the observation time, t, uncertainty, Δt. The H value in the relations has an action dimension and is dependent upon the precision of the rod's clocks and its mass.

Taking into account the fact that in Russia the recommendations formulated by meteorologists for replacing the terms "error" with the term "uncertainty" are not compulsory in nature, and bearing in mind that these recommendations are still in the discussion stage [1–2], we used both terms in the work at hand.

We will take error to mean the inaccuracy of measurement results that is due to purely metrological causes. Increasing an instrument's measurement accuracy can reduce the error. The absolute error of the distance between two points can serve as an example of an error.

We will call the inaccuracy of measurement results that cannot be eliminated by means of increasing measurement instrument accuracy, and which may be due to terminological, conceptual, or linguistic causes, the uncertainty of physical values. The uncertainty of the distance between two spheres that are located close to one another can serve as an example of this uncertainty. This distance remains uncertain with an accuracy of up to the dimensions of the spheres, even in the presence of ideal measurement accuracy, while the thing that remains unclear is what is meant by the sought distance – the distance between the centres of mass of the spheres, the distance between their geometric centres, the distance between the closest points of the spheres, or something else. Construing uncertainty in this manner is mentioned in extant literature, albeit in general terms. In this vein, for example, taking into account the fact that it is only possible to specify the location of a spatially extended body by determining the position of a single solitary point that belongs to it with some degree of uncertainty, the uncertainty of the position of a sphere determined by the position of its centre, which equals the radius of this sphere, is written about in reference [3].

The uncertainty of specifying the moment in time of a short-term process that does not occur instantaneously, but rather occupies a certain finite, let's say, very short time interval, can serve as another example of an uncertainty of this type. This uncertainty can be regarded as equal to half the duration of the process, if the moment in time of its transit is called the moment into which the middle of the process falls. These coordinate and time uncertainties, Δx and Δt, are precisely the ones that play a part in the uncertainty relations we derived.

E.1 Single-Time and Single-Coordinate Data. Determining the Velocity of a Rod Based on the Readings of the Synchronized Clocks Appurtenant to It

We will visualize a thin rod of proper length, L, at two points, a and b, on which synchronously running clocks, A and B, are installed at a distance of d from one another.

Let's say that clocks A and B, like the clocks appurtenant to the K^0 reference system, where the rod is at rest, show this system's time; i.e. the readings of clocks A and B are always in agreement with the readings of the K^0 system's clocks. The length, d, of section ab, which is located between points a and b, may be equal to or less than the rod length, L; i.e. the condition $L \geq d$ holds true in the general case. If $L > d$, then the rod will look something like this:

———-A————————-B————-

Here, the characters A and B conditionally designate clock A and clock B, while the broken line shows the body of the rod.

If the distance, d, is equal to the rod length, L, i.e. if $L = d$, clocks A and B are then found at the ends of the rod.

We will call the arrangement that consists of this rod and the two running clocks, A and B, situated on it rod, R; i.e. clocks A and B will be regarded as integral parts of rod R, and we will treat the readings of clocks A and B as characteristic attributes of rod R.

Let's say that rod R, which is positioned parallel to the X^0 axis of inertial system K^0, is at rest relative to this system and moves at a constant velocity, V', along the X' axis of another reference system, K', remaining parallel to the X' axis and its direction of movement (the X^0 and X' axes of the K^0 and K' systems slide along one another over the course of their relative motion). We will call this rod motion longitudinal motion (with respect to its orientation in space), and it alone will be referred to in the future.

Pursuant to inverse Lorentz transformations, at a moment in time of t' of the K' system, the readings, $\tau_{A,t'}$ and $\tau_{B,t'}$, of clocks A and B, which are in agreement with the readings of the K^0 system's clocks, are determined by the relations

$$\tau_{A,t'} = \frac{t' + \frac{V'}{c^2} x'_{A,t'}}{\sqrt{1 - \left(\frac{V'}{c}\right)^2}} \text{ and } \tau_{B,t'} = \frac{t' + \frac{V'}{c^2} x'_{B,t'}}{\sqrt{1 - \left(\frac{V'}{c}\right)^2}},$$

where $x'_{A,t'}$ and $x'_{B,t'}$ are the coordinates of clocks A and B of rod R within system K' at a moment in tie of t', while c is the speed of light in a vacuum.

As follows from the relations presented above, the difference in the readings of clocks A and B of rod R at a moment in time of t' for system K' equals

$$\tau_{B,t'} - \tau_{A,t'} = \frac{(x'_{B,t'} - x'_{A,t'})V'}{c^2\sqrt{1 - \left(\frac{V'}{c}\right)^2}}. \tag{E.1}$$

Introducing the following notation for the purpose of saving writing space

$$U' = \frac{V'}{\sqrt{1 - \left(\frac{V'}{c}\right)^2}}, \tag{E.2}$$

then from formula (E.1), with allowance for notation (E.2), we obtain:

$$U' = \frac{c^2(\tau_{B,t'} - \tau_{A,t'})}{x'_{B,t'} - x'_{A,t'}}. \tag{E.3}$$

Thus, having the data on the coordinates and readings of clocks A and B at a moment in time of t', formula (E.3) can be used to find the value of U', and this value can in turn be used to find the velocity, V', of rod R in system K'. We will call the data that characterize the object's spatially distributed elements, but that relate to one and the same moment in time, t', single-time data.

In addition to the possibility of determining the rod's velocity using single-time data, the possibility also exists of determining the velocity, V', of rod R in system K' using single-coordinate data. We will call the data successively recorded at different moments in time, but at one and the same point (with one and the same coordinate), which characterize the elements of a spatially extended object at the times that they are located at this (or near this) point, single-coordinate data. Data of this type include, for example, the $\tau_{A,x'}$ and $\tau_{B,x'}$ readings of clocks A and B of rod R that are successively recorded at moments in time of $t'_{A,x'}$ and $t'_{B,x'}$ at a point in system K' with a coordinate of x', past which it is moving within this reference system.

According to inverse Lorentz transformations, the single-coordinate $\tau_{A,x'}$ and $\tau_{B,x'}$ readings of clocks A and B are connected to the moments in time, $t'_{A,x'}$ and $t'_{B,x'}$, during which clocks A and B, at a point with a coordinate of x', show the relations

$$\tau_{A,x'} = \frac{t'_{A,x'} + \frac{V'}{c^2}x'}{\sqrt{1-\left(\frac{V'}{c}\right)^2}} \text{ and } \tau_{B,x'} = \frac{t'_{B,x'} + \frac{V'}{c^2}x'}{\sqrt{1-\left(\frac{V'}{c}\right)^2}}.$$

As follows from the relations presented above, the difference in the readings of clocks A and B of rod R at a point with a K' system coordinate of x' equals

$$\tau_{B,x'} - \tau_{A,x'} = \frac{t'_{B,x'} - t'_{A,x'}}{\sqrt{1-\left(\frac{V'}{c}\right)^2}}. \tag{E.4}$$

Introducing the notation $\Gamma' = 1/\sqrt{1-(V'/c)^2}$, then from formula (E.4), we obtain

$$\Gamma' = \frac{\tau_{B,x'} - \tau_{A,x'}}{t'_{B,x'} - t'_{A,x'}}, \tag{E.5}$$

from which the velocity, V', of rod R can be learned as necessary.

The determination of velocity using single-coordinate data is interesting in that it does not require distance measurements, but rather is based on the measurement of the $\tau_{B,x'} - \tau_{A,x'}$ and $t'_{B,x'} - t'_{A,x'}$ time intervals.

E.2 Relation of Rod Velocity and Coordinate Uncertainties Calculated Using Single-Time Data

In talking about the determination of the velocity of rod R using the difference in the readings of clocks A and B, we tacitly proceeded on the basis of the fact that clocks A and B of rod R run ideally; i.e. the readings of clocks A and B are in absolutely precise agreement with the ideally accurate readings of the K^0 system clocks. We speculate that a certain maximum absolute error exists in the time readings of each of clocks A and B, which are appurtenant to rod R. We will designate the absolute error of these clocks as $\Delta\tau$. Here, adhering to the generally accepted assumption, we will presume that all the clocks appurtenant to any inertial reference system, including the K^0 system, run with ideal accuracy.

The existence of an absolute error, $\Delta\tau$, in the readings of each of clocks A and B means that, at a given moment in time, t^0, within the K^0 system, the readings of each of these clocks of rod R can differ from the readings of the K^0 system clocks located close to them and running with ideal accuracy by a value that does not exceed $\Delta\tau$. In this regard, the remarks made above concerning the agreement of the reading of the K^0 system clocks with those of clocks A and B must be interpreted with allowance for the finite accuracy of the latter.

If the rate of the K' system clocks, the measured $x'_{B,t'} - x'_{A,t'}$ value, and the speed of light c value are as accurate as desired, the error in the U' value calculated using formula (E.3) will then be solely due to the existence of an absolute error, $\Delta(\tau_{B,t'} - \tau_{A,t'})$, in the difference of the $\tau_{B,t'} - \tau_{A,t'}$ readings of clocks A and B.

In this instance, the $\Delta U'$ error, with allowance for formula (E.3), is expressed by the equality

$$\Delta U' = \frac{c^2 \Delta(\tau_{B,t'} - \tau_{A,t'})}{x'_{B,t'} - x'_{A,t'}} \tag{E.6}$$

In instances when the maximum absolute error of the difference in the readings of clocks A and B consists of the $\Delta\tau$ errors of each of these clocks, i.e. when $\Delta(\tau_{B,t'} - \tau_{A,t'}) = 2\Delta\tau$, it follows from equality (E.6) that

$$(x'_{B,t'} - x'_{A,t'})\Delta U' = 2c^2\Delta\tau. \qquad \text{(E.7)}$$

We note that the $\Delta\tau$ error is not dependent upon reference system selection, since it consists of the maximum possible difference between the readings of each of clocks A and B and the readings of the K^0 system clocks located close to them, where rod R is at rest. It is clear that this difference is not dependent upon the reference system within which it is picked up.

We will now imagine that, in addition to the condition of the single-time nature of the readings of clocks A and B within system K', the requirement of the single-coordinate nature of the specification of the location where rod R is situated at a moment in time of t' is satisfied. Let's say the essence of this requirement consists of using a single solitary x'_R coordinate to specify the location of the projection of rod R on the X' axis. This requirement can be satisfied if the rod length is ignored and it is regarded as a point. But if, due to the necessity of taking the property of the rod's spatial extent into account, it is impossible to ignore its length, the requirement of the single-coordinate nature of the specification of the position of rod R can only be satisfied in part. For example, the coordinate of any point appurtenant to rod R can be specified as its x'_R coordinate and a reference to its uncertainty can accompany the specification of this coordinate. In particular, the coordinate of the geometric centre of rod R, or the coordinate of its centre of mass, can serve as the coordinate of this point. In such cases, the distance from the point with a coordinate of x'_R to the point of the rod's projection on the X' axis farthest away from it can be regarded as the uncertainty, $\Delta x'_R$, of the x'_R coordinate.

When the position of rod R is specified in this manner, the x'_R coordinate indicates the location of one of a set of points of the projection of rod R that lies on the X' axis. If we give this point preference for one reason or another, the $\Delta x'_R$ uncertainty will then determine a range of point coordinates that, in the presence of other considerations, could also be regarded at the point coordinates of rod R.

For example, if the coordinate, x'_{ab}, of the geometric centre (the midpoint) of the ab section of rod R parallel to the X' axis is selected as this section's coordinate, the uncertainty, $\Delta x'_{ab}$, of the x'_{ab} coordinate of the rod's ab section can by definition be regarded as a value equal to half the length of the section of the moving (within the K' system) rod.

Since the length, $d' = d\sqrt{1-(V'/c)^2}$, of the rod's moving ab section within the K' system equals $x'_{B,t'} - x'_{A,t'}$, the uncertainty, $\Delta x'_{ab}$, of the x'_{ab} coordinate of section ab of rod R then equals $1/2\ (x'_{B,t'} - x'_{A,t'})$. Thus, formula (E.7) can be presented in the form

$$\Delta U'\Delta x'_{ab} = c^2\Delta\tau. \tag{E.8}$$

Within the framework we have provided when introducing the determination of uncertainty, the $\Delta x'_{ab}$ value is absolutely an uncertainty and not an error, since it is dependent upon the length of the rod's ab section and cannot be reduced by means of increasing measurement accuracy.

Because $L \geq d$ in the general case, the $\Delta x'$ uncertainty of the x' coordinate of rod R within the arbitrary positioning of clocks A and B thereon ($\Delta x' = 1/2L'$) can then generally both equal and exceed the $\Delta x'_{ab}$ value. And since the relation $\Delta x' \geq \Delta x'_{ab}$ holds true, then in the general case of the arbitrary positioning of the clocks on the rod, formula (E.8) takes the form:

$$\Delta U'\Delta x' \geq c^2\Delta\tau. \tag{E.9}$$

Relation (E.9) pertains to the general case of the arbitrary positioning of the clocks on the rod and makes the transition to the equality $\Delta U'\Delta x' = c^2\Delta\tau$ in the special and most favourable case (as far as the determination of the U' value using single-time data) of clock positioning on the ends of the rod ($\Delta x' = \Delta x'_{ab}$).

The product of the error, $\Delta U'$, of the U' value and the uncertainty, $\Delta x'$, of the x' coordinate of rod R during an instantaneous observation will only be dependent upon the error in the readings of clocks A and B of rod R. Therefore, relation (E.9) remains unchanged in any inertial reference system and can be written for an arbitrary system in the form:

$$\Delta U_x\Delta x \geq c^2\Delta\tau \tag{E.10}$$

If the mass of rod R with clocks A and B is known without a doubt and equals M_R (here and further on, the concepts of Lorentz-invariant mass [4] will be used), relation (E.10) can be transformed into the relation $M_R \Delta U_x \Delta x \geq M_R c^2 \Delta\tau$ by multiplying its left-hand and right-hand members times M_R, whence, taking into account the fact that $M_R U_x = p_x$, we obtain:

$$\Delta p_x \Delta x \geq M_R c^2 \Delta\tau. \tag{E.11}$$

Introducing the notation

$$H = M_R c^2 \Delta\tau, \tag{E.12}$$

then from formula (E.11), we obtain

$$\Delta p_x \Delta x \geq H. \tag{E.13}$$

The $\Delta\tau$ error of clocks A and B of rod R is one of the internal parameters of rod R and is not a part of the parameters of the measuring devices that are external relative to rod R. Therefore, the ΔU_x and Δp_x errors in relations (E.10) and (E.13) can tentatively be referred to as external (relative to rod R) uncertainties. The internal properties of rod R itself (a salient feature of clocks A and B appurtenant to it) determine the external uncertainties ΔU_x and Δp_x, and in the presence of a specific Δx uncertainty, they cannot be eliminated by means of increasing the accuracy of the measuring instruments located beyond rod R.

E.3 Relation of Rod Energy and Time Uncertainties Calculated Using Single-Coordinate Data

The error, $\Delta\Gamma'$, of the Γ' value can be derived from formula (E.5). Only the $\tau_{B,x'} - \tau_{A,x'}$ value error due to the $\Delta\tau$ error in the readings of clocks A and B determines the $\Delta\Gamma'$ error of the Γ' value, since the $t'_{B,x'} - t'_{A,x'}$ time interval at point x', according to our initial assumption, can be measured by the K' system clocks with an ideal accuracy. Therefore, from formula (E.5), we obtain:

$$\Delta\Gamma' = \frac{\Delta(\tau_{B,x'} - \tau_{A,x'})}{t'_{B,x'} - t'_{A,x'}},$$

or with allowance for the fact that $\Delta(\tau_{B,x'} - \tau_{A,x'}) = 2\Delta\tau$,

$$\Delta\Gamma' = \frac{2\Delta\tau}{t'_{B,x'} - t'_{A,x'}},$$

whence it follows that

$$\Delta\Gamma'(t'_{B,x'} - t'_{A,x'}) = 2\Delta\tau. \tag{E.14}$$

We will now assume that, in addition to the condition of the single-coordinate nature of the observation clock readings, the requirement of the single-time nature of the specification of the observation time of the *ab* section of rod R, moving past point x', is imposed. Let's say the requirement consists of using a single solitary moment in time, $t'_{ab,x'}$, to specify the section *ab* observation time at point x'.

Since the observation at a point with a coordinate of x' is carried out over a time interval of $t'_{B,x'} - t'_{A,x'}$, it is then only possible to approximately specify the observation time by indicating, e.g. the moment in time, $t'_{ab,x'}$, of the middle of the time interval, $t'_{B,x'} - t'_{A,x'}$, that goes towards observation. Here, the uncertainty, $\Delta t'_{ab,x'}$, of the moment in the observation time that equals half the $t'_{B,x'} - t'_{A,x'}$ observation time can be specified; i.e. it can be assumed that $\Delta t'_{ab,x'} = 1/2(t'_{B,x'} - t'_{A,x'})$. Then formula (E.14) can be presented in the form

$$\Delta\Gamma'\Delta t'_{ab,x'} = \Delta\tau. \tag{E.15}$$

For the arbitrary positioning of the clocks on the rod, the $\Delta t'_{x'}$ observation time for the entire rod, R, with a length of L proves to be greater than the observation time, $\Delta t'_{ab,x'}$, for the *ab* rod section; i.e. the condition $\Delta t'_{x'} \geq \Delta t'_{ab,x'}$ is satisfied, as a result of which it follows from formula (E.15) that

$$\Delta\Gamma'\Delta t'_{x'} \geq \Delta\tau. \tag{E.16}$$

During an accurate observation, the product of the uncertainty, $\Delta t'_{x'}$, of a moment in time, $t'_{x'}$, of the observation of rod R and the $\Delta\Gamma'$ error of the Γ' value will only be dependent upon the $\Delta\tau$ error of the readings of clocks A and B of rod R. Therefore, relation (E.16) remains unchanged within any inertial reference system and can be written for an arbitrary system in the form

$$\Delta\Gamma\Delta t_x \geq \Delta\tau. \tag{E.17}$$

Multiplying the left-hand and right-hand members of relation (E.17) times $M_R c^2$, bearing in mind that $\Delta\Gamma M_R c^2 = \Delta E$, and using notation (E.12), we obtain

$$\Delta E\,\Delta t_x \geq H. \qquad \text{(E.18)}$$

If the $\Delta\tau$ error is regarded as a rod parameter that is not a part of the measuring equipment parameters, the $\Delta\Gamma$ and ΔE values can then be referred to as uncertainties.

E.4 A Physical Clock. Relation of the Impulse and Coordinate Uncertainties of a Spatially Extended Body

We will assume that each of clocks A and B discretely changes its reading with frequency, ν. The maximum absolute $\Delta\tau$ error in the readings of the moments in time of each of the clocks will equal $1/\nu$. We will also assume that this change in the clock readings occurs synchronously; i.e. not only are the clock readings at a given moment in time synchronized, but also the times of the change in readings.

In this instance, the maximum error of the difference in the readings of clocks A and B will equal not the sum of the two $\Delta\tau$ errors, but rather a single $\Delta\tau$ error. Then relations (E.13) and (E.18), deduced from the condition of the equality of the clock A and B difference error to the two $\Delta\tau$ errors, take the form:

$$\Delta p_x \Delta x \geq H/2 \qquad \text{(E.19)}$$

and

$$\Delta E\,\Delta t_x \geq H/2. \qquad \text{(E.20)}$$

Let's imagine that each of clocks A and B consists of two components, one of which, being artificial ("manmade"), performs the function of a display and provides discrete time readings, changing them in a keeping with external signals, while the other one – we will call it a physical clock – generates these signals in a natural, easy manner and controls the change in the display's readings.

For the sake of clarity, we will visualize the second part of the physical clock as a piece of radioactive material with a long half-life (the material's radiant power can be regarded as constant for a sufficiently long time frame).

If the display reacts to a specific portion, ε, of the absorbed gamma-radiation energy physical clock's material by changing its readings, then in the presence of a material radiant power that equals P, the frequency, ν, of the change in the display readings will equal P/ε.

We will call the ε portion of the absorbed energy that leads to a change in the display's reading the energy of the perceived (by the display) physical clock signals.

We will assume that, regardless of the quantity of the material of the physical clock, the display absorbs all of the energy radiated, and that each ε portion of the energy absorbed performs the function of a reading change signal that is perceived by the display. The frequency, ν, of the physical clock materials signals perceived by the display, and accordingly the frequency of the change in the physical clock's readings, will then be proportional to the quantity of the physical clock's material. This means that if the frequency of the perceived signals equals ν_0 in the presence of a physical clock material unit mass of m_0, it will then equal $M_0\nu_0/m_0$ in the presence of a mass of M_0; i.e.

$$\nu = M_0 \frac{\nu_0}{m_0}, \tag{E.21}$$

Since the maximum absolute $\Delta\tau$ error of the readings of each of the clocks equals $1/\nu$, the following expression can then be derived from equality (E.21):

$$\Delta\tau = \frac{m_0}{M_0\nu_0}. \tag{E.22}$$

Taking the mass of the rod without the clocks to be negligible as compared to the mass of the physical clock material concentrated in clocks A and B, or assuming that the rod R consists entirely of the physical clock's material, the $2M_0$ value (since the mass of the material in both clocks equals $2M_0$) can be set to equal the mass, M_R, of rod R with clocks A and B. Taking this into account, it follows from formulas (E.12) and (E.22) that

$$H = \frac{2c^2 m_0}{\nu_0} \tag{E.23}$$

The H value is dependent upon physical clock type and display sensitivity, and is not a physical constant. The H becomes different when the display's sensitivity changes, or when the physical clock's radioactive material is replaced with a radioactive material that has a different radiant power.

In the context of the physical clock, relations (E.19) and (E.20) pertain to the case of the arbitrary distribution of the physical clock's material along the rod and make the transition to equalities in the special case when the physical clock's material is concentrated at the ends of rod R.

At first glance, the relations derived only hold true for the techniques of instantaneous and point observations of an object, and the uncertainties going into these relations consist exclusively of the uncertainties inherent in these techniques. In actuality, however, it is impossible to measure even the constant velocity of this rod R, equipped with clocks A and B, with absolute accuracy using conventional methods (based on the path traversed and the time), if the word combination "rod R with clocks A and B" is taken to mean a specific object, the complement of characteristic traits of which includes the difference in the readings of clocks A and B (or the events that characterize this object). In such cases, the velocity and the values derived from it (the impulse and energy) prove to be tied not to an extraneous reference system, but rather to this object's characteristics traits, for example, to time or event marks [5].

In order to grasp the essence of the latter comment, we will address the concepts of relativity and uncertainty.

It was shown in reference [5] that many of the questions arising in the quagmire of physical relativism are erased if attention is directed to the presence of uncertainty in physical relativity. Rather than talking about the relativity of the velocity of a body and about the fact that the velocity of a body without specifying a reference system makes no physical sense, it is proposed in reference [5] that the term "uncertain velocity" be used with respect to the irrelative velocity of a body. In this case, the velocity of a body that is not tied to a specific reference system should be referred to as uncertain velocity. The uncertainty of the irrelative velocity of a body equals the c constant. It can be said of the irrelative velocity of this body, for example, that it equals zero, while its uncertainty equals c. This specification of velocity does not differ qualitatively in any way from the conventional specification of a velocity value supplemented by the specification of its error.

We draw attention to two methods that make it possible to avoid the uncertainty of the velocity of a body and to give it certainty.

The first generally accepted method consists of specifying the reference system within which the velocity of a body is examined. After the reference system is specified, the velocity of this body becomes certain.

The second method, despite its obviousness, is scarcely mentioned at all in physics. This method consists of individualizing an object, which in turn consists of describing it in greater detail, and under certain conditions, it is capable of replacing the selection of a reference system.

In the example we examined, the object with respect to which relations (E.19) and (E.20) hold true is not a rod, R, equipped with clocks A and B, but rather a rod, R, equipped with clocks A and B, "and possessing a predetermined difference, $\tau_B - \tau_A$, in the readings, τ_A and τ_B, of clocks A and B."

If one and the same rod, R, equipped with clocks A and B, is by definition regarded as a set of more concrete sub-objects, each of which has an inherent difference, $\tau_B - \tau_A$, in the readings of clocks A and B, then these different objects will possess different velocities. The possibility of thus dividing an object into more concrete sub-objects eludes relativists, since it puts an end to the objective nature of physical relativism. Relativists are incapable of understanding something that was clear to Heracles, who saw the difference between one and the same river and particular concrete "sub-objects" of this river that differed from one another. At the same time, relativists themselves note that one and the same object differs in different reference systems due to the relativity of its single-time nature; i.e. it breaks down into sub-objects of sorts. It is strange that they do not comprehend the fact that, having described a sub-object, its velocity can often be determined based on "external appearance" without specifying a reference system.

For example, in the case we examined, a sub-object such as a rod, R, that has a specific difference, $\tau_{B,t'} - \tau_{A,t'}$, in the readings, $\tau_{A,t'}$

and $\tau_{B,t'}$, of clocks A and B, moves at a specific longitudinal velocity, V_x. When the clocks run with ideal accuracy, the V_x velocity value is functionally dependent upon the $\tau_{B,t'} - \tau_{A,t'}$ difference value. A sub-object such as a rod, R, with a difference, $\tau_{B,t'} - \tau_{A,t'}$, in the readings, $\tau_{A,t'}$ and $\tau_{B,t'}$, of clocks A and B that equals zero is at rest. Its longitudinal velocity, V_x, equals zero and cannot differ from zero in any reference system, since sub-objects with a difference, $\tau_{B,t'} - \tau_{A,t'}$, in the readings of clocks A and B that equals zero do not exist in any reference system where this rod moves longitudinally. This rod, R, with clocks A and B, has different longitudinal velocities in different reference systems, but the subject rod, R, with a difference, $\tau_{B,t'} - \tau_{A,t'}$, in the readings of clocks A and B that equals zero, being a concrete sub-object, only has a longitudinal velocity that equals zero. This fact is entirely independent of the manner in which the velocity is measured - based on the clock readings or using the path and distance traversed.

In addition to this, if clocks A and B provide discrete readings, the $\tau_{B,t'} - \tau_{A,t'}$ difference will also have a certain discreteness. Here, the longitudinal velocity value, V_x, will have an uncertainty of ΔV_x. In this case, a rod, R, with a given difference in the readings of clocks A and B, being a sub-object, may be found in a certain set of reference systems in a state of motion at various longitudinal velocities that differ from one another by a magnitude that does not exceed a certain value, which is the ΔV_x uncertainty. This uncertainty in accuracy equals the uncertainty that occurs during the single-time determination of the V_x velocity of a given sub-object.

If clocks A and B do not run, but rather stand still, continuously indicating an identical time, then the ΔV_x uncertainty of the longitudinal velocity of rod, R, with a difference of $\tau_{B,t'} - \tau_{A,t'}$ in the readings of clocks A and B that equals zero, will equal the speed of light, c. Such a rod, R, with a difference of $\tau_{B,t'} - \tau_{A,t'}$ in the readings of clocks A and B that equals zero, can be found in all reference systems and possesses any velocity over a range of zero to the speed of light, c.

E.5 Conclusion

The purpose of the work at hand consisted not of finding areas of common interest between Lorentzian physics and quantum mechanics, but rather of demonstrating the existence of general physical uncertainty relations in the physics of the macrocosm. The general physical relations derived are externally reminiscent of the known uncertainty relations of quantum mechanics; however, the physical essence of the values that go into the relations and that contain the relations themselves are different than those in quantum mechanics. In relation (E.13), the Δx uncertainty is taken to mean the uncertainty in specifying the position of a projection with a length of Δx of a spatially extended object using a single solitary coordinate, x, while in the Heisenberg relation, Δx is a probabilistic characteristic of the position of a micro-particle described by the root-mean-square deviation from the mean value.

The Δp_x value in relation (E.13) is the p_x impulse error, which in the presence of a predetermined Δx uncertainty, generally speaking, can be reduced by using another type of physical clock, while Δp_x in the Heisenberg relation is an uncertainty that it is fundamentally impossible to reduce in the presence of a predetermined Δx value.

Different meanings are placed on the H and h values.

The H value in relations (E.13) and (E.18)–(E.20) is dependent upon display sensitivity and physical clock type. If the sensitivity of the display is changed or the physical clock's radioactive material is replaced with a radioactive material that has a different mass radiation frequency, the H value will be different. However, the fundamental Planck's constant, h, is unrelated to the physical properties of any specific material. The fact that these relations prove to be connected to the Heisenberg relation, not only externally, but also internally, seems especially strange.

First, there is the question of the minimum value, $H_{\min}$, of an H parameter with an action dimension. Can one assume that this value may be as small as desired in the macrocosm and may correspond, for example, to the condition $H_{\min} \ll h$?

And second, relations (E.19) and (E.20) are formally transformed into the relations

$$\Delta p_x \Delta x \geq h/2 \text{ and } \Delta E \Delta t_x \geq h/2$$

via the simple substitution of the energy, $h\nu_0$, of a photon with a frequency of ν_0 in place of the unit energy, m_0c^2, of the physical clock in formula (E.23); i.e. by taking the unit mass of a photon, the ν_0 frequency of which numerically equals the ν_0 frequency of the signals of a hypothetical change in the clock readings as the physical clock. It would be more correct to proceed on the basis of the concept of Lorentz-invariant mass and the equality of photon mass to zero, then a pair of photons coming from opposite directions with equal energies of $E = h\nu$ and with a resultant impulse, P, that equals zero, should be regarded as a physical clock of a unit mass of m_0. Since the unit mass, m_0, of this clock equals $\sqrt{(\frac{2E}{c^2})^2 - (\frac{P}{c})^2} = 2E/c^2$, while the ν_0 summation frequency of the electromagnetic oscillations equals 2ν, then taking the ν_0 frequency of the unit mass physical clock to equal the 2ν summation frequency of the photon pair, it follows from formula (E.23) that $H = H_{\min} = h$.

Another approach is also possible. In considering formula (E.23), if it is assumed that a certain $H_{\min}$ value exists that is common to all types of physical clocks, then $2m_0c^2/\nu_0 = H_{\min}$. The latter equality only occurs when $m_0c^2 = 1/2H_{\min}\nu_0$. If the lowest energy of a hypothetic signal of a change in clock readings is taken as m_0c^2, then $H_{\min}$ must be equal to the Planck's constant h, since the $1/2\ h\nu_0$ value is the minimum possible energy of the zero-point oscillations of an oscillator with a frequency of ν_0.

The relations expressed by formula (E.23), if we move to them from relations (E.19) and (E.20), do not reflect the statistical nature of the generation of the clock reading change signals; thus, when using this approach, reference can only be made to the order of the parameter present in the right-hand member of the relations and not its precise value.

References

1. Tarbeyev Yu. V., Slayev V. A., Chunovkina A. G. Problems with using the international guide to the expression of uncertainty in measurement in Russia. *Izmeritelnaya Tekhnika* [*Measurement Techniques*] **1**, 1997, 69–72.

2. uropean Society for Analytical Chemistry/International Cooperation on Traceability in Analytical Chemistry (EURACHEM/CITAC) Guide Entitled *A Quantitative Description of Uncertainty in Analytical Measurement* (2nd Ed., 2000) – Translated From the English. D. I. Mendeleyev Russian National Metrology Research Institute (RNMRI), St. Petersburg, 2002.
3. Sukhanov A. D., Golubeva O. N. *Lectures on Quantum Physics*. Vysshaya Shkola [Higher School] Publishing House, Moscow, 2006, p. 54.
4. Okun L. B. *Advances in the Physical Sciences (APS)*, No. 178, 2008, pp. 541–555.
5. Matveev V. N. *Into the Third Millennium Without Physical Relativity*. CheRo Publishing House, Moscow, 2000.

References

Many of the following texts have been chosen because they are *not* arcane academic texts gathering dust in university libraries being fairly readily available. In this age of the Internet, book stores such as Amazon, Barnes, and Noble, and Abe Books provide a repository of original texts or digitized data (e-books) from around the globe. The scientific pool of knowledge contained within them is thus widespread and not esoteric or restricted in nature. It has been the author's pleasure to read these texts or parts thereof in order to gain an overview of the world's scientific knowledge at this point in time at the start of the 21st century. The polymath student will doubtless enjoy the task of delving into the layers of intellectual wealth across the ancient and more recent generations, a form of archeology no less rewarding than hen's teeth.

General History of Science

Barnett, J. E., *Time's Pendulum: The Quest to Capture Time, From Sundials to Atomic Clocks*, Plenum, Cambridge, 1998.

Bodanis, D., *$E = mc^2$: A Biography of the World's Most Famous Equation*, Macmillan, London, 2000.

Braun, J., *Music in Ancient Israel/Palestine: Archaeological, Written, and Comparative Sources*, Eerdmans, 2002.

Clegg, B., *Light Years*, Piatkus, London, 2001.

Cropper, W. H., *Great Physicists*, Oxford University Press, New York, 2001.

Ferris, T., *The Red Limit: The Search For the Edge of the Universe*, Perennial, New York, 2002.

Gibbon, J., *Science a History, 1543–2001*, Penguin, London, 2002.

Gleick, J., *Genius: Richard Feynman and Modern Physics*, Abacus, London, 2000.

Hamilton, J. D., *Faraday: The Life*, HarperCollins, London, 2003.

Hawking, S., *On the Shoulders of Giants: The Great Works of Physics and Astronomy*, Running Press, Philadelphia, 2002.

Hunt, V. H., *Origins In Acoustics: The Science of Sound From Antiquity to the Age of Newton*, Yale University Press, New Haven, CT, 1978.

Rhodes, R., *The Making of the Atomic Bomb*, Simon and Schuster, 1988.

Rouse Ball, W. W., *A Short Account of the History of Mathematics*, Dover Publications, 4th Ed., New York, 2003.

Thorne, K. S., *Black Holes and Time Warps: Einstein's Outrageous Legacy*, W.W. Norton and Company, New York, 1995.

Whittaker, E., *A History of the Theories of Aether and Electricity*, Vol. 1, Philosophical Library, New York, 1954.

Whittaker, E., *A History of the Theories of Aether and Electricity*, Vol. 2, Harper & Brothers, New York, 1960.

Williams, H. S., *The Story of Nineteenth Century Science*, Harper & Brothers, 1904.

Yaghjian, A., *Relativistic Dynamics of A Charged Sphere: Updating the Lorentz-Abraham Model*, Lecture Notes in Physics, 2nd Ed., Springer, 2005.

Mathematics

Jancewicz B, *Multivectors and Clifford Algebra in Electrodynamics*, World Scientific, Singapore, 1988.

Misra, J. C., *Biomathematics: Modeling and Simulation*, World Scientific, Singapore, 2006.

Schutz, B. F., *Geometrical Methods of Mathematical Physics*, Cambridge University Press, Cambridge, 1980.

Thomas Jr. G. B., *Calculus and Analytic Geometry*, World Student Series, 3rd Ed., Addison-Wesley, 1966.

Electromagnetics

Barut, A. O., *Electrodynamics and Classical Theory of Fields and Particles*, Dover, New York, 1980.

Harrington, R. F., *Time Harmonic Electromagnetic Fields*, Mc-Graw-Hill, New York, 1961.

Kraus, J. D., *Antennas*, McGraw-Hill Electrical and Electronic Engineering Series, McGraw-Hill, 1950.

Lehnert, B., Roy, S., *Extended Electromagnetic Theory: Space-Charge in Vacuo and the Rest Mass of the Photon*, World Scientific Series in Contemporary Chemical Physics, World Scientific, Singapore, 1999.

Melia, F., *Electrodynamics, Chicago Lectures in Physics*, University of Chicago Press, London, 2001.

Jackson, J. D., *Classical Electrodynamics*, 3rd Ed., John Wiley & Sons, New York, 1999.

Von Hippel, A., *Dielectrics and Waves*, John Wiley & Sons, New York, 1962.

Maxwell, J. C., *Treatise on Electricity and Magnetism*, Vol. 1 and 2, 3rd Ed., Publications, 1954.

Schwartz, M., *Principles of Electrodynamics*, Dover, New York, 1987.

Quantum Theory

Condon, E. U., Odabasi, H., *Atomic Structure*, Cambridge University Press, Cambridge, 1980.

Heisenberg, W., *The Physical Principles of the Quantum Theory*, Dover, New York, 1949.

Heitler, W., *The Quantum Theory of Radiation*, Dover, New York, 1984.

House, J. E., *Fundamentals of Quantum Mechanics*, Academic Press, London, 1998.

McMahon, D., *Quantum Field Theory Demystified*, McGraw-Hill, 2008.

Miller, A. I., *Early Quantum Electrodynamics: A Sourcebook*, Cambridge University Press, Cambridge, 1995.

Srednicki, M., *Quantum Field Theory*, Cambridge University Press, Cambridge, 2007.

Physics

Yariv, A., *Quantum Electronics*, 3rd Ed., Wiley & Sons, 1988.

Ashcroft, N. W., Mermin, N. D., *Solid State Physics*, Holt-Saunders International Editions, 1976.

Einstein, A., *Relativity, the Special and General Theory*, Three Rivers Press, New York, 1961.

Einstein, A., The Meaning of Relativity, in *Including the Relativistic Theory of the Non-Symmetric Field*, 5th Ed., MJF Books, New York, 1984.

Einstein, A., Relativity, the special and general theory, in *Albert Einstein in His Own Words*, Portland House, New York, 2000.

Fowler, C. M. R., *The Solid Earth: An Introduction to Global Geophysics*, Cambridge University Press, Cambridge, 1990.

Hawking, S., *A Brief History of Time*, Bantam, London, 1988
Henson, R., *The Rough Guide to Weather*, Rough Guides, 2002.

Kittel, C., *Quantum Theory of Solids*, 2nd Ed., Wiley, 1987.

Maxwell, J. C., *Matter and Motion*, Great Minds Series Reprint Ed., Prometheus Books, 2002.

Pauling, L., *The Nature of the Chemical Bond and the Structure of Molecules and Crystals, an Introduction to Modern Structural Chemistry*, 3rd Ed., George Fisher Baker Non-Resident Lecture, Cornell University Press, 1960.

Pierce, A. D., *Acoustics: An Introduction to Its Physical Principles and Applications*, American Institute of Physics, 1989.

Stevens, K. W. H., *Magnetic Ions in Crystals*, Princeton University Press, 1997.

Rindler, W., *Relativity: Special, General, and Cosmological*, Oxford University Press, Oxford, 2004.

Nuclear and Particle Physics

Cottingham, W. N., Greenwood, D. A., *An Introduction to the Standard Model of Particle Physics*, Cambridge University Press, Cambridge, 1998.

Heisenberg, W., *Introduction to the Unified Field Theory of Elementary Particles*, Interscience, New York, 1966.

Krane, K. S., *Introductory Nuclear Physics*, John Wiley & Sons, New York, 1988.

Perkins, D. H., *Introduction to High Energy Physics*, 4th Ed., Cambridge University Press, Cambridge, 2003.

Watson, A., *The Quantum Quark*, Cambridge University Press, Cambridge, 2004.

Astronomy and Cosmology

Adams, F., Laughlin, G., *The Five Ages of the Universe: Inside the Physics of Eternity*, Touchstone, New York, 1999.

Binney, J., Tremaine, S., *Galactic Dynamics*, Princeton Series in Astrophysics, Princeton University Press, Princeton, NJ, 1987.

Mather, J. C., Boslough, J., *The Very First Light*, Penguin Books, London, 1998.

North, J. D., *The Measure of the Universe*, Clarendon Press, Oxford, 1960.

Biology and Bioelectromagnetics

Chang, R., *Physical Chemistry with Applications to Biological Systems*, 2nd Ed., Prentice Hall College, 1977.

Gagliardi, L. J., Electrostatic force in prometaphase, metaphase, and anaphase-A chromosome motions, *Endogeneous Physical Fields in Biology*, URSI International Symposium July 1–3, Prague, Czech Republic, 2002.

Haeger, K., *The Illustrated History of Surgery: Revised and Updated by Sir Roy Calne FRS*, Harold Starke, London, 2000.

Ling, G. N., *Life at the Cell and Below-Cell Level*, Pacific Press, Melville, New York, 2001.

Popp, F. A., About the coherence of biophotons, microscopic quantum coherence, Proceedings of an International Conference, World Scientific, River Edge, NJ, 1999.

Beloussov, L. V., Voeikov, V. L., Martynyuk, V. S., *Biophotonics and Coherent Systems in Biology*, Springer, 2006.

Gilbert, F. G., *Developmental Biology*, 9th Ed., Sinauer Associates, 2010.

Ahmed, A. J., Alizad, A., *Biomedical Applications of Vibration and Acoustics in Therapy, Bioeffect and Modeling*, ASME, New York, 2008.

Ho Mae-Wan, F. A Popp., Warnke, U., *Bioelectrodynamics and Biocommunication*, World Scientific, 1994.

Polk, C., *CRC Handbook of Biological Effects of Electromagnetic Fields*, CRC Press, 1986.

Rosch, P. J., Markov, M. S., *Bioelectromagnetic Medicine*, Marcel Dekker, 2004.

Binhi, V. N., *Magnetobiology: Underlying Physical Problems*, Academic Press, 2002.

Lund, E. J., *Bioelectric Fields and Growth*, University of Texas, 1947.

Zewail, A., Ed., *Physical Biology: From Atoms to Medicine*, Imperial College Press, 2008.

Self-Field Theory

Fleming, A. H. J., Self-Field Theory: A New Mathematical Description of Physics, Int. Scientific Congress 2010, Fundamental Problems of Natural Sciences and Eng., St. Petersburg, Russia July 26–30, 2010.

Fleming, A. H. J., Self-Field Theory: Are Cosmological and Biological Evolution Linked? Int. Scientific Congress 2010, Fundamental Problems of Natural Sciences and Eng., St. Petersburg, Russia July 26–30, 2010.

Fleming, A. H. J., Self-Field Theory: Biodiversity May Be a Resonance Process, PIERS-10, Cambridge, MA, June, 2010.

Fleming, A. H. J., Self-Field Theory: A Possible Gravitational Structure for Galaxies, PIERS-10, Cambridge, MA, June, 2010.

Fleming, A. H. J., Self-Field Theory: Biophotons and EPR, PIERS-10, Cambridge, MA, June, 2010.

Fleming, A. H. J., Self-Field Theory: New Photonic Insights, PIERS-10, Xi'an, China, March, 2010.

Fleming, A. H. J., Analytic Estimate of the Mass of the Photon, PIERS-09, Moscow, Russia, September, 2009.

Fleming, A. H. J., The Role of Endogenous and Exogenous E-fields in Metaphase, PIERS-09, Beijing, China, March, 2009.

Fleming, A. H. J., Self-Field Theory: The Spectroscopy of the Photon, BEMS-30, San Diego, June 2008.

Fleming, A. H. J., Self-Field Theory: Analytic Spectroscopy of the Ordinary Photon, 2nd International Conference on Electromagnetic Fields, Health and Environment, Wroclaw, Poland, September 10–12, 2007.

Fleming, A. H. J., Bauer, E. B., Bergeron, R., Dahle, M., Enge, J., Acoustic/Magnetic Field Assisted Perfusion, 2nd International Conference on Electromagnetic Fields, Health and Environment Wroclaw, Poland, September 10–12, 2007.

Fleming, A. H. J., Self-Field Theory: The Mass of The Photon and the Fine Structure Constant, BEMS-29, Kanazawa, Japan, June 2007.

Fleming, A. H. J., Self-Field Theory: Mathematical Perspective, BEMS-28, Cancun, Mexico, June 2006.

Fleming, A. H. J., Electromagnetic self-field theory and its application to the hydrogen atom, *Physics Essays*, **(18)**3, 2005.

Fleming, A. H. J., Bauer, E. B., Case Study: Part A Electromagnetic Self Field Theory, BEMS-27, Dublin, June 2005.

Bauer, E. B., Cooper, K., Jenks, P., Fleming, A. H. J., Case Study: Part B, The Effects of Acoustic Frequencies on Core Tendon Lesions of the Thoroughbred Racehorse, BEMS-27, Dublin, June 2005.

Fleming, A. H. J., Colorio, E. B. (nee Bauer), The Spectroscopy of the EM Field: A Predicted Photon Chemistry, 3rd International Workshop on Biological Effects of EM Fields, Kos, Greece, October 2004.

Fleming, A. H. J., Colorio, E. B. (nee Bauer), A Predicted Photon Chemistry, BEMS-26, Washington DC, June 2004.

Other References

Fleming, A. H. J., Towards Computational Methods for Studying Cellular Effects Due to EM Field, Applied Computational Electromagnetics Conference, Naval Postgraduate College, Monterey, California, March 15–19, 1999.

Fleming, A. H. J., Farrell, P. M., A Model for Ionic Diffusion in a Liquid Due to Static and Time-varying E- and B-Fields, Sixteenth Annual Meeting Bioelectromagnetics Society, Copenhagen, June 12–17, 1994.

Fleming, A. H. J., Joyner, K. H. Estimates of absorption by the embryo and foetus during pregnancy due to radio-frequency radiation exposure, *Health Physics*, 149–159, 1992.

Fleming, A. H. J., Joyner, K. H. (eds.), Special issue on bioelectromagnetic computations, *Applied Computational Electromagnetic Society Journal* **7**(2), 1992.

Fleming, A. H. J., Joyner, K. H. Radio-frequency radiation exposure and pregnancy, Joint Annual Conference Australian Radiation Protection Society and Australasian College of Physical Scientists and Engineers in Medicine, Adelaide, 24–28 September, 1990.

Fleming, A. H. J. A finite element method for composite scatterers, Chapter 2 in *Finite Element and Finite Difference Methods in Electromagnetic Scattering* (Morgan, M. A., Ed.), Progress in Electromagnetics Research, Vol. 2, New York, Elsevier, pp. 69–112, 1990.

Hocking, B., Joyner, K. H., Fleming, A. H. J., Implanted medical devices in workers exposed to radio-frequency radiation, *Scandinavian Journal of Work Environment and Health*, **17**, 1–6, 1990.

Fleming, A. H. J., Joyner, K. H., Moment method analysis of radiation and scattering by thin wires in an infinite conducting medium, Special Issue *ACES Journal on Electromagnetic Code Validation*, 51–74, 1989 (special issue).

Index

Color Insert

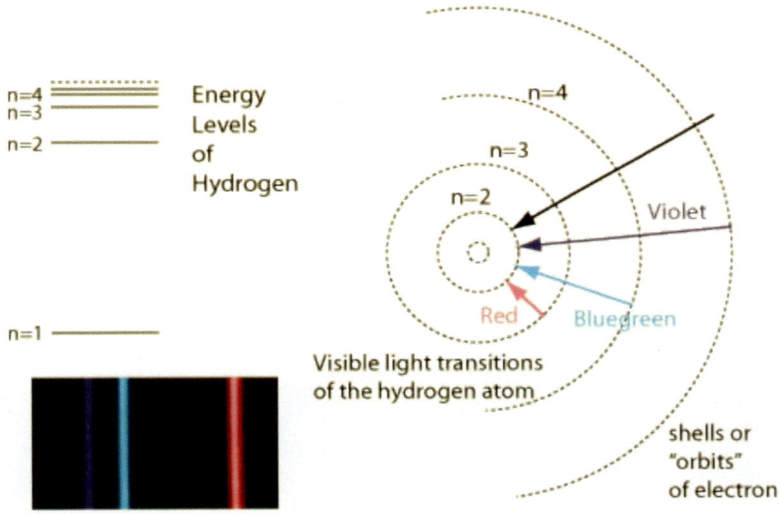

Figure 1.3. Quantized levels of the atom. (Credit: Wikimedia.)

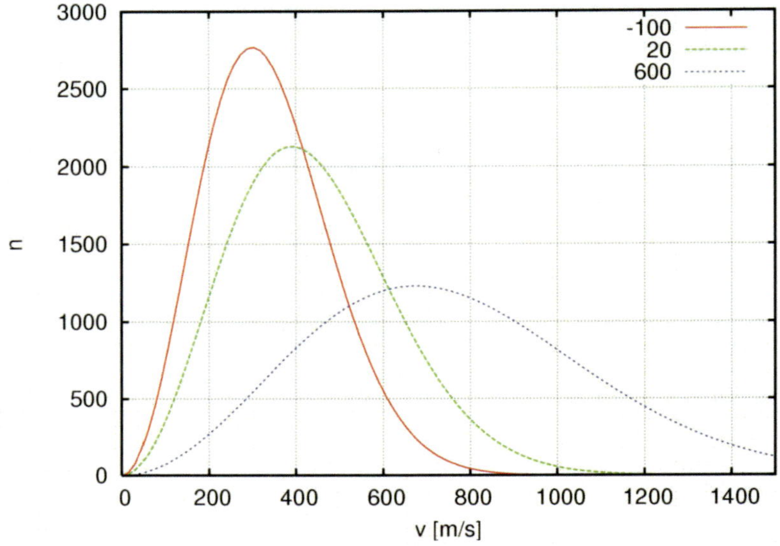

Figure 1.8. Maxwell–Boltzmann distribution for thermodynamics. (Credit: Wikimedia.)

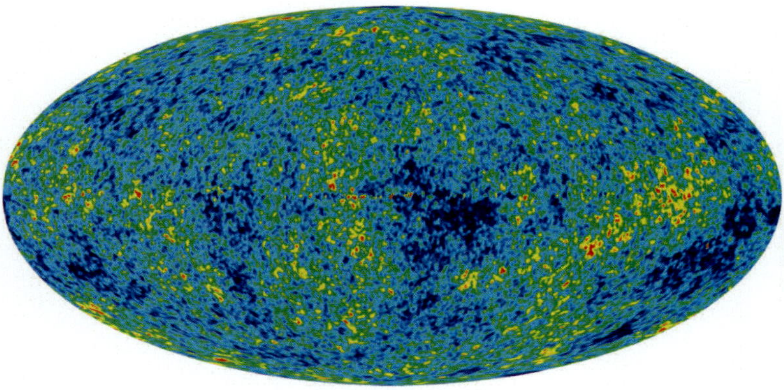

Figure 1.10. Cosmic microwave background temperature fluctuations over the full sky. (Credit: Wikimedia.)

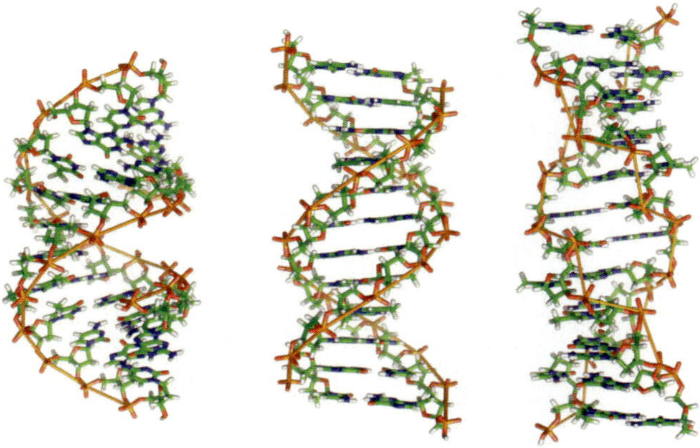

Figure 1.11. Piecewise bi-spinorial structure of the double-helix of the DNA. (Credit: Wikimedia.)

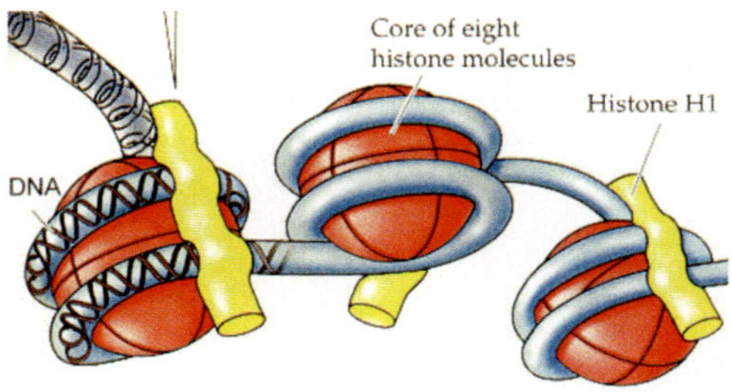

Figure 1.12. Histones seen as "beads-on-a-string" within the coiled double helix of the DNA. (Credit: Wikimedia.)

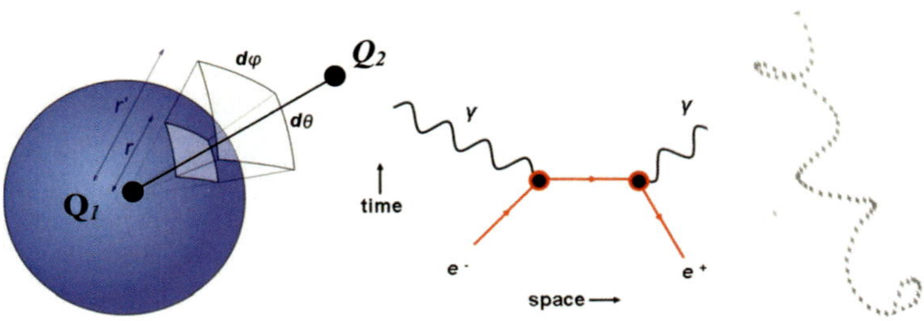

Figure 2.3. Field forms: (a) classical electromagnetic, (b) QFT, and (c) SFT. (Credit: Wikimedia.)

Figure 2.11. Earth's rotation is the sum of two dielectromagnetic effects. (Credit: Wikemedia.)

Figure 3.1. Stonehenge. (Source: http://www.english-heritage.org.uk/server/show/nav.16465/chosenImageId/12)

Figure 3.2. Torre dei Venti. (Source: http://asv.vatican.va/it/arch/torre.htm)

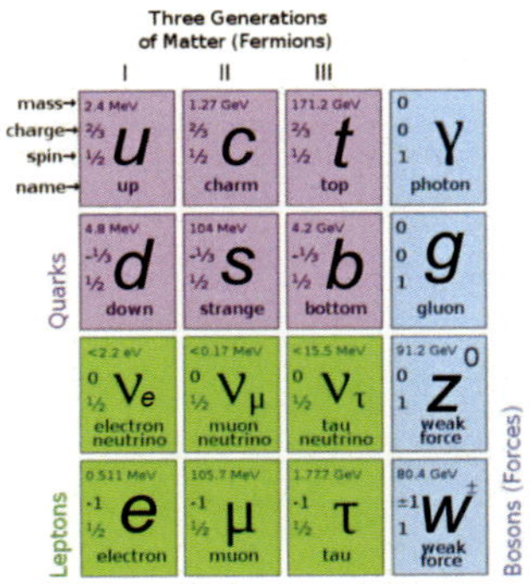

Figure 3.4. Standard model of elementary particles. (Credit: Wikimedia.)

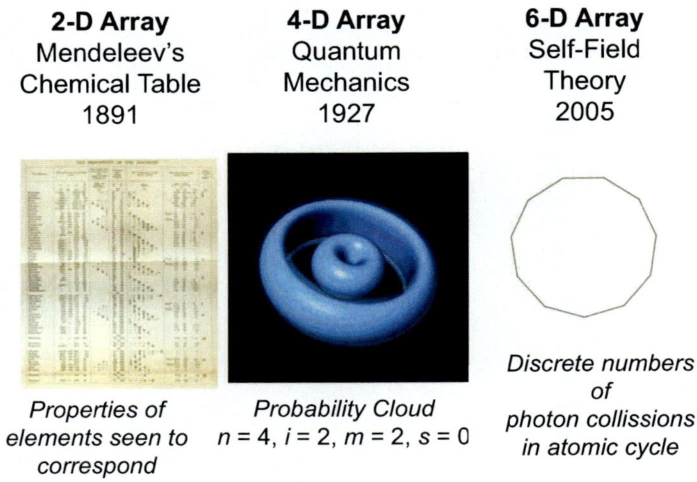

Figure 3.9. Historical perspective of the two new SFT quantum numbers associated with the resonance of the photon in its role as the binding energy in atoms and molecules. (Credit: Wikimedia).

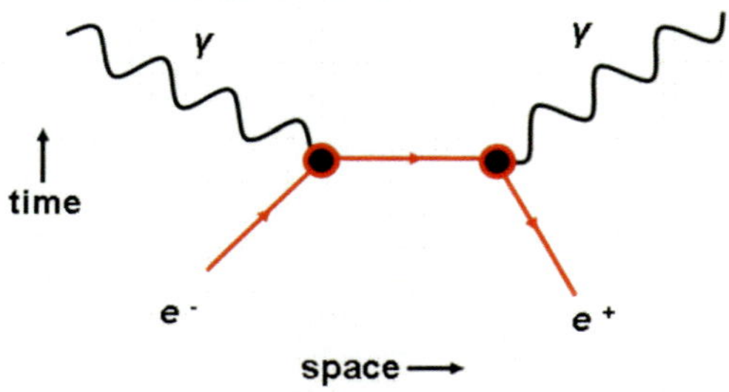

Figure 3.12. Anihilation of electron positron pair. (Credit: Wikimedia.)

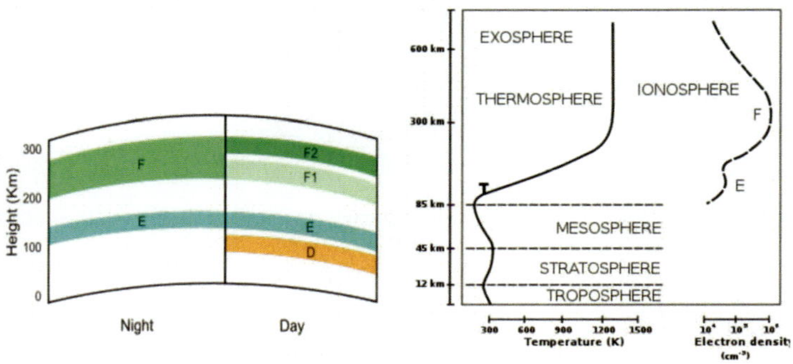

Figure 3.14. Layered structure of atmosphere and ionosphere.

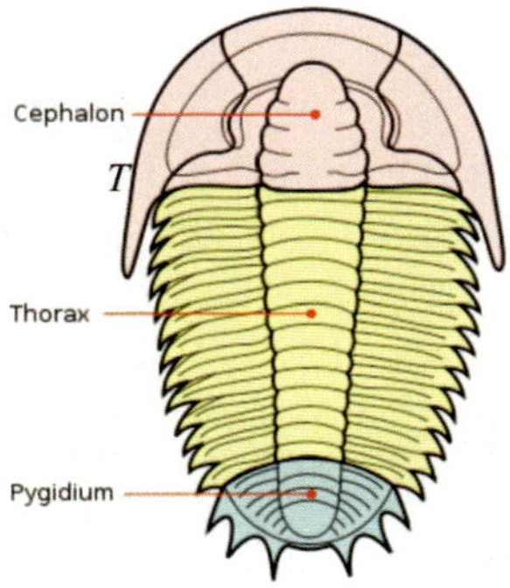

Figure 4.2. The trilobite body; cephalon, thorax, and pygidium. (Credit: Wikipedia.)

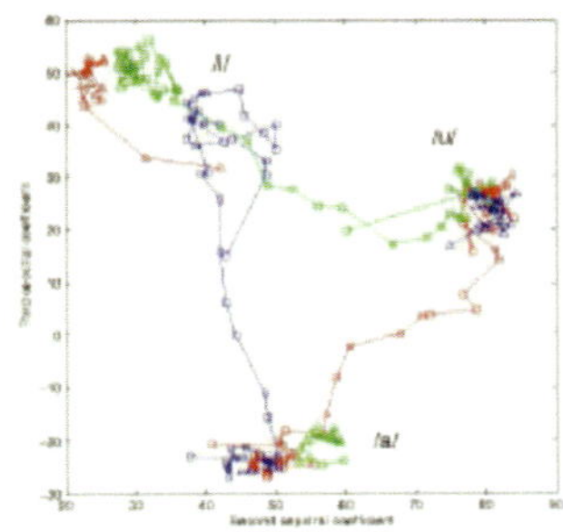

Figure 4.3. Trajectory representation derived from recorded acoustic data of Dutch speech. (Source: Zuidema, W., de Boer, B., The Evolution of Combinatorial Phonology, University of Amsterdam Institute for Logic, Language and Computation, 2006.)

Figure 4.4. Jadeware with description of Qigong from Warring States Period 476–221 BC. (Source: http://tcm.chinese.cn/en/article/2009-08/24/content_11271. htm)

Figure 4.6. Didgeredoo. (Source: http://medgadget.com/archives/2007/01/do_you_have_sle.html)

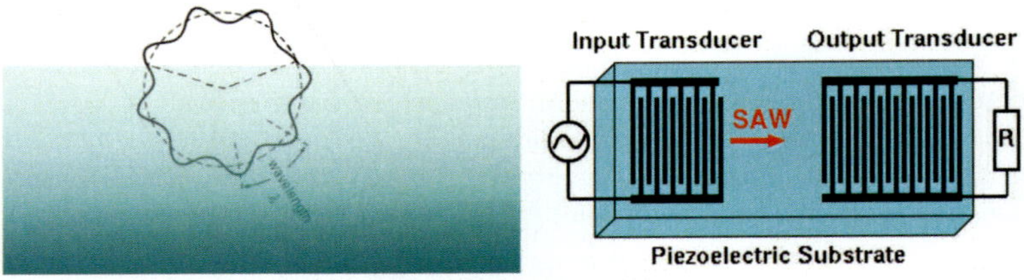

Figure 4.7. (a) Surface acoustics wave (b) SAW electronic component. (Credit: Wikipedia.)

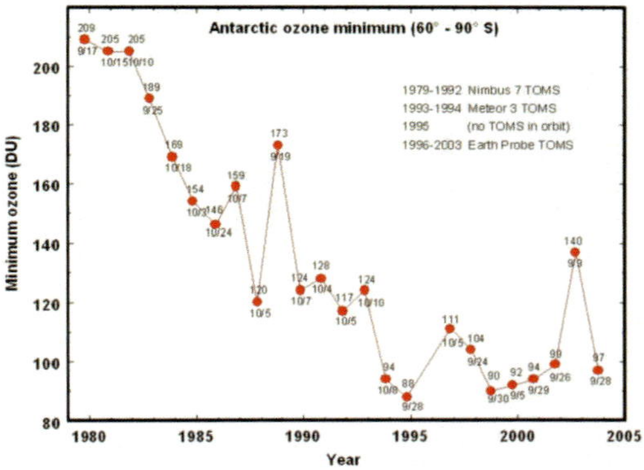

Figure 4.10. Ozone depletion, 1979–2004. (Source: http:/toms.gsfc.nasa.gov/multi/multi.html)

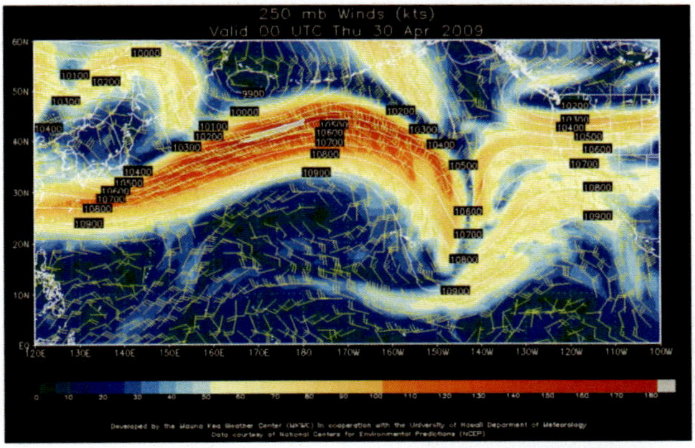

Figure 4.11. Wind forecast for North Pacific, April 30, 2009. (Source: Http://mkwc.ifa.hawali.edu/)

Figure 4.13. According to SFT lightning, wind, and thunder are forms of electric, magnetic, and acoustic energy. (Source: www.stormscapedarwin.com)

Figure 4.14. Galactic dynamics: typical sun orbiting around the centre of a galaxy. (Source: http://kusmos.phsx.ku.edu/~melott/mmcr.htm)

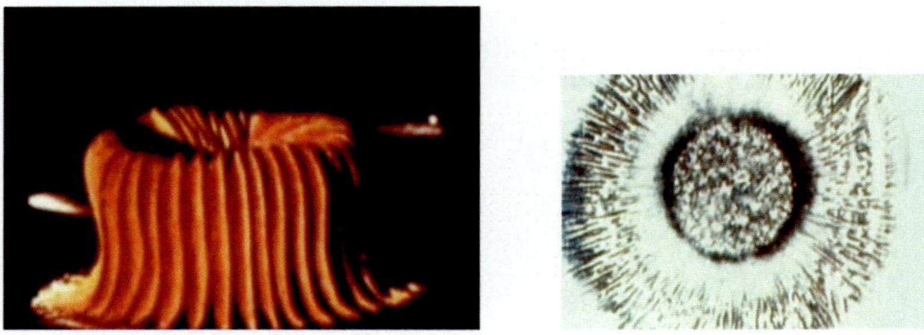

(a) (b)

Figure 4.16. Exposure to sound (a) liquid plastic appears as a cake (b) iron filings as an eye. (Source: http://www.unitedearth.com.au/sound.html)

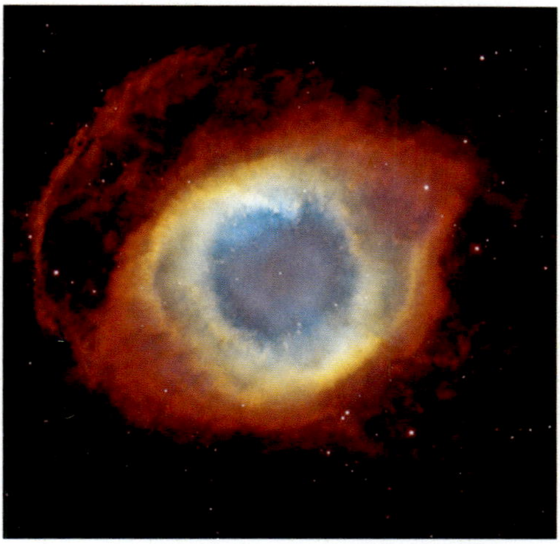

Figure 4.17. Helix nebula, NGC 7293, sometimes known as "The Eye of God." (Credit: Wikimedia.)

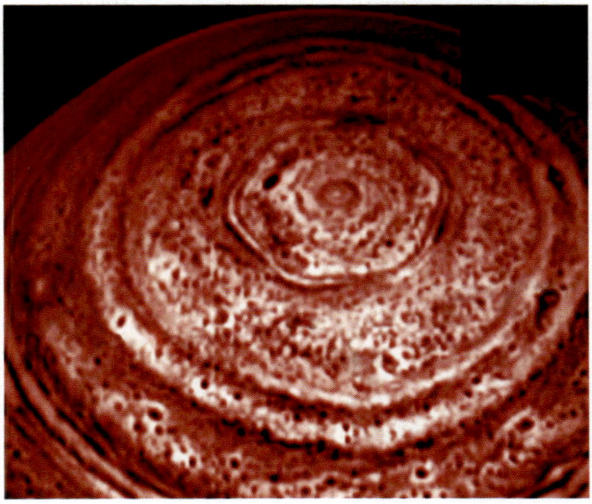

Figure 4.18. Hexagon structure surrounding north pole of Saturn of side about. (Credit: NASA/JPL/University of Arizona.)

Figure 4.19. Induction coil used in 1800s to show Faraday's law of induction $\varepsilon = |\mathrm{d}\phi_B/\mathrm{d}t|$. (Credit: Wikipedia.)

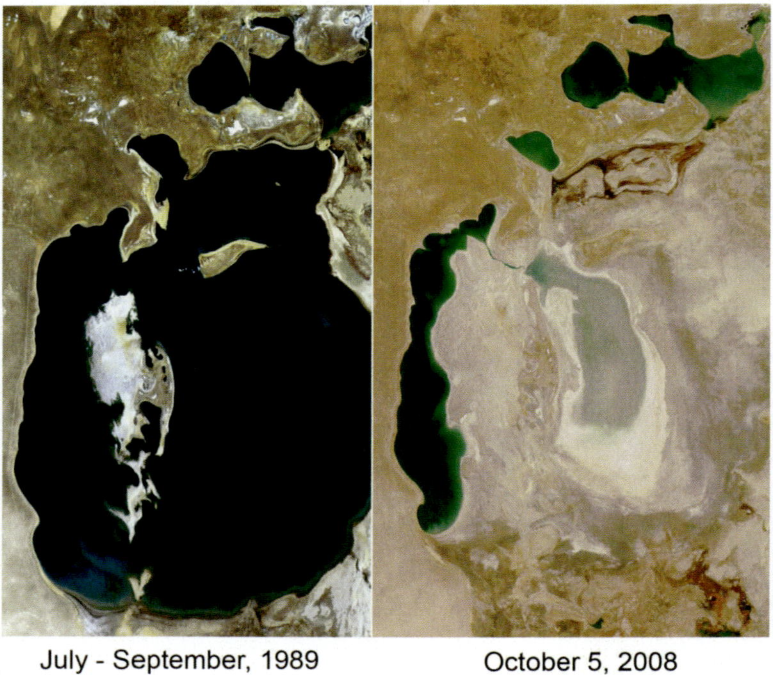

Figure 5.1. Drying of the Aral Sea due to Soviet irrigation diversion. (Source: Wikipedia.)

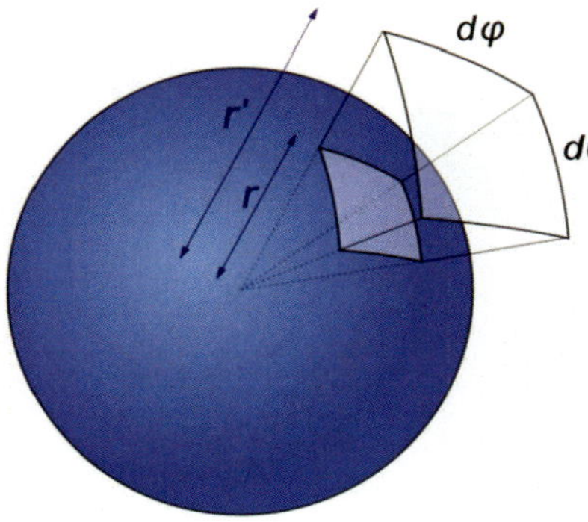

Figure A.1. In SFT a finite number of boson streams transit across a small but finite solid angle as they moves between particles, not over all solid angles. (Credit: Wikipedia.)

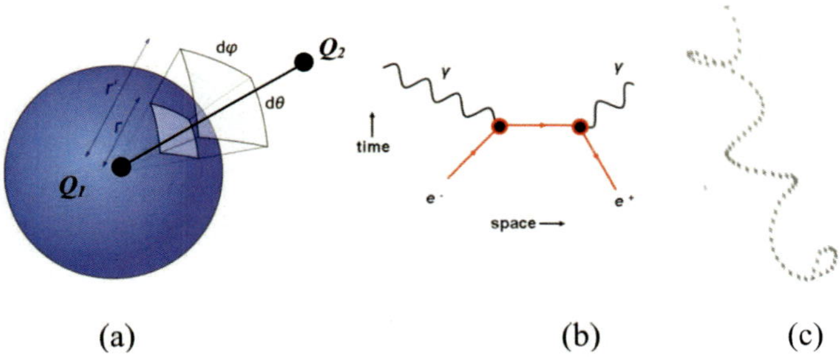

Figure C.1. Field forms (a) CEM, (b) QFT, (c) SFT.